LASER APPLICATIONS IN ENVIRONMENTAL MONITORING

LASER APPLICATIONS IN ENVIRONMENTAL MONITORING

LUCA FIORANI AND FRANCESCO COLAO
EDITORS

Nova Science Publishers, Inc.
New York

For permission to use material from this book please contact us:
Telephone 631-231-7269; Fax 631-231-8175
Web Site: http://www.novapublishers.com

NOTICE TO THE READER

The Publisher has taken reasonable care in the preparation of this book, but makes no expressed or implied warranty of any kind and assumes no responsibility for any errors or omissions. No liability is assumed for incidental or consequential damages in connection with or arising out of information contained in this book. The Publisher shall not be liable for any special, consequential, or exemplary damages resulting, in whole or in part, from the readers' use of, or reliance upon, this material. Any parts of this book based on government reports are so indicated and copyright is claimed for those parts to the extent applicable to compilations of such works.

Independent verification should be sought for any data, advice or recommendations contained in this book. In addition, no responsibility is assumed by the publisher for any injury and/or damage to persons or property arising from any methods, products, instructions, ideas or otherwise contained in this publication.

This publication is designed to provide accurate and authoritative information with regard to the subject matter covered herein. It is sold with the clear understanding that the Publisher is not engaged in rendering legal or any other professional services. If legal or any other expert assistance is required, the services of a competent person should be sought. FROM A DECLARATION OF PARTICIPANTS JOINTLY ADOPTED BY A COMMITTEE OF THE AMERICAN BAR ASSOCIATION AND A COMMITTEE OF PUBLISHERS.

LIBRARY OF CONGRESS CATALOGING-IN-PUBLICATION DATA

Laser applications in environmental monitoring / Luca Fiorani and Francesco Colao(editors).
     p. cm.
Includes bibliographical references and index.
ISBN 978-1-60456-249-1(hardcover)
    1.   Environmental   monitoring.2.Lasers—Scientific   applications.I.Fiorani,Luca.II. Colao,Francesco.
TD193.L376 2008
628—dc22
              2007050817

Published by Nova Science Publishers, Inc. ✦ New York

CONTENTS

PREFACE

Earth, water, air and fire: According to the ancient Greek philosopher Empedocles, they are the elements of all matter. Indeed, the three main spheres of our natural environment are mainly made of earth (lithosphere), water (hydrosphere) and air (atmosphere). This book ideally completes the Empedocles' set of elements by reporting on a recent addition to earth, water and air, i.e., a modern fire constituted by the laser source.

Lithosphere, hydrosphere and atmosphere play a major role in sustaining life on our planet. All the biogeochemical cycles take place in them or are part of the matter exchange among them. A thorough understanding of such global processes requires powerful tools.

Environmental monitoring investigates natural processes, anthropogenic effects and their interactions. It is often performed by passive remote sensing, using the Sun as a light source, or active remote sensing, being assisted by artificial light sources. In both cases, the matter of lithosphere, hydrosphere and atmosphere interacts with the energy of the source. The information on earth, water and air is contained in the radiation coming back from those elements. The passive technique relies on observing the planet, i.e., on measuring the sunlight backscattered by its surface at different wavelengths. It suits especially satellite-borne sensors. The active method is usually based on laser sources. It is applied mainly to ground-based, air-, ship- and, more recently, satellite-borne sensors.

The revolutionary capabilities offered to environmental monitoring by the laser discovery have been firstly understood and exploited in the atmospheric field. The measurement, up to many kilometers of height, of density, temperature and humidity of air, the detection of trace gases, the study of clouds, the observation of stratospheric aerosols, the probing of high atmosphere and the monitoring of pollutants are some examples among the possible atmospheric applications of lasers.

Lasers have been successfully used also in the hydrosphere for bathymetric surveys in shallow waters, turbidity measurements, pollution detection - especially in case of oil spill - and phytoplankton mappings.

As far as the lithosphere is concerned, laser employment has been limited by the lack of light propagation in soil. In this case, the more relevant environmental applications of laser are related to the volumetric characterization of hollow spaces beneath the terrestrial surface by laser range-finder and the detection of soil pollutants by laser-induced breakdown spectroscopy.

The purpose of this book is twofold: on one hand, the interested student is introduced to laser application in environmental monitoring; on the other hand, the professional familiar with the topic is given information some current investigations. The first aim is achieved by introducing the measurement principles. The second one is pursued by describing in detail the most recent results obtained by specialists in the field.

This book is divided in to three chapters, reporting on laser application to air, water and soil. Their content has been written by specialists working in the respective field for decades, both in private companies and governmental institutions in the United States and Europe.

The first chapter is a small encyclopedia of atmospheric laser radars (lidars). After an introduction to the lidar principle, the lidar equation is presented, and bias and noise sources are described in detail. Also, the necessary information on atmospheric structure is given. All kinds of known radiation-matter processes relevant to atmospheric sounding are clearly discussed. Three subchapters accurately report on the main techniques in atmospheric lidar: elastic backscatter lidar, Raman backscatter lidar and differential absorption lidar. Many interesting applications by different researchers spanning the globe and focused on different atmospheric regions exemplify the content of the chapter. Eventually, the present status and future perspectives of satellite-borne lidar are illustrated.

Laser application in water monitoring is discussed in the second chapter. It is divided in to two subchapters, the first one devoted to lidar bathymetry and the second one to laser-induced fluorescence. Lidar bathymetry is a promising technique for coastal zone monitoring. This approach is described from the point of views of historical development, scientific basis, technological solutions, data processing and actual applications. Two case studies end the subchapter (the second one is a very interesting survey of a river basin). Laser-induced fluorescence is a very sensitive analysis tool of dissolved and particulate matter in natural waters, e.g., chromophoric dissolved organic matter and chlorophyll-a. It makes possible the continuous retrieval of concentration profiles of such components, thus supporting our understanding of the biogeochemical cycles taking place in seas and oceans. Another interesting application of laser-induced fluorescence is the detection of oil spills. The description of an advanced system used in Antarctic waters, a submarine lidar fluorosensor payload, concludes the subchapter.

The book ends with the third chapter, devoted to two important laser applications in soil monitoring, i.e., soil analysis by laser-induced breakdown spectroscopy and three-dimensional scan of underground cavities, treated in two separate subchapters. Laser-induced breakdown spectroscopy, after the pioneering years, has become a mature analytical tool for elemental concentration determination. In this case, the laser beam interacts strongly with the matter target to generate a plasma plume. After an accurate treatment of principles of operation, typical instrumentations, plasma physics, qualitative and quantitative measurements, soils applications are introduced. The case studies encompasses analysis of soil and sediments, and even extraterrestrial exploration. The three-dimensional scan of underground cavities is the conceptually simplest use of laser radar, i.e., rangefinding. Nevertheless, an accurate volumetric characterization of buried spaces is a demanding technological challenge. After an introduction to rangefinding, with special emphasis on laser techniques, and interferometry, the main methods are introduced: time of flight, optical triangulation and amplitude modulation. An example of a promising application of this latter method, the imaging topological radar (ITR), is then described, showing its capability of 3D-model reconstruction. Eventually, the deployment of ITR in a prehistoric cave is reported.

We wish the reader the same pleasure and excitement we had in editing this book.

1. LASER APPLICATIONS IN AIR MONITORING

In: Laser Applications in Environmental Monitoring
Editors: L. Fiorani and F. Colao

ISBN 978-1-60456-249-1
© 2008 Nova Science Publishers, Inc.

Chapter 1

ATMOSPHERIC LIDARS

Valentin Mitev

CSEM - Centre Suisse d'Electronique et de Microtechnique SA,
rue Jaquet-Droz 1, Case postale, CH2002 Neuchâtel, Switzerland
E-mail: Valentin.Mitev@csem.ch, tel.: +41 – 32-889 88 13

ABSTRACT

The Atmospheric lidars use the propagation and the scattering of the laser beam in the air to deduce information about the measured parameters. The lidars are remote sensing instruments measuring the range resolved parameters of the atmosphere in a continuous way. The introduction and practical application of this type of instrument are based on the progress in lasers, the optical detectors and the optical materials, and the signal processing electronics. After almost four decades of development, the lidar is presently a widely used and indispensable instrument for atmospheric measurements from surface, aircraft and satellites. The lidar provides valuable measurements of the aerosol backscatter and extinction; it allows the possibility to distinguish between droplets and ice-crystals in clouds; it monitors the diurnal cycle of the mixing layer, the atmospheric temperature, the content of water vapour, ozone and various pollution gases, and wind. The atmospheric altitudes of lidar measurements are from just above the surface to the top of the mesosphere. The investigated phenomena encompass a wide range of time-scale, from fast mixing processes to climatology of important atmospheric parameters. The purpose of this article is to introduce the reader to the basic principles used in the most widespread atmospheric lidars, the way the information about the atmospheric parameters is retrieved from the detected lidar signal, as well as to the main requirements in the realisations. The article also lists examples for representative lidar realisations and applications.

1. INTRODUCTION

The atmospheric lidars operate by transmitting a light beam and processing the backscattered response. The name "lidar" is an abbreviation of "Light Detection and Ranging". By its principle and by its abbreviation, the lidar is the analog of the radar in the optical domain. Although the idea for lidar was introduced before the invention of the laser [*Middleton and Spilhaus 1953*], it was this source of powerful, low divergent and monochromatic optical pulse that made it practical [*Smullin and Fiocco 1962, Fiocco and Grams 1964*]. The further lidar development is based on the progress in photonics, i.e., lasers, optical materials and designs, detectors, as well as signal processing electronics.

The lidar applies the same principle as radar, i.e., propagation and scattering of electromagnetic waves, but in the optical domain, where the wavelength is much shorter. The sizes of the atmospheric molecules and aerosol particles are comparable to the optical wavelength, so they are efficiently scattering. The shorter wavelength in the optical domain makes the lidar transmitter and receiver smaller than the radar antenna and with a narrower field of view.

Compared to the other atmospheric instruments, both *in situ* and remote sensors, there are several advantages that make the atmospheric lidar attractive.

The lidar is a remote sensing instrument. That is, to obtain information about the atmospheric volume, we do not need to be in contact with it. The lidar is an active optical instrument. It may provide range-resolved information independently on the direction and the daily cycle of the natural light sources. The possibility for continuous measurements is another advantage, as it provides the dynamics of the atmospheric parameter. Anyway, with all these advantages, the atmospheric lidar does not replace the other instruments, but complements them.

The scope of this article is to introduce the reader to the basic physical principles of the atmospheric lidar: the atmospheric parameters that can be retrieved from the lidar signal and what the main requirements are of the lidar hardware in the various lidar types. The quoted examples show milestones in the lidar development and its present state-of-the-art. It is, indeed, not possible to cover all used phenomena and all existing variants of lidars, or its numerous realizations. So the efforts are limited to the most widely-used elastic backscatter lidar, the Raman lidar and the differential-absorption lidar (DIAL). There are other lidar types, both technically advanced and with high information potential: the fluorescence lidar, high-resolution lidar, and Doppler wind lidar, but with deployment still limited in scale. Those interested more in the lidar technique will find further reading in [*Weitkamp 2005, Hinkley 1976, Measures 1984*], as well as in the specialised papers quoted further.

2. THE LIDAR SET-UP AND THE LIDAR PRINCIPLE

2.1. The Set-Up

Figure 1 illustrates the lidar principle. A laser beam is directed into the atmosphere, in such a way, that its propagation is confined inside the field-of-view of the lidar receiver telescope. Sometimes a beam expander is used to decrease the divergence of the transmitted laser beam. The receiver collects the part of the light scattered in a backward direction during the laser beam propagation in the atmosphere. All optical and laser subsystems of the lidar are assembled together on the same mechanical support frame, keeping the co-alignment between the transmitted laser beam and the telescope field of view. The typical divergence of the transmitted laser beam is in the order of 0.1-0.5mrad. The field of view of the receiver varies typically from parts of mrad to order of mrad, respectively.

The partial and the full overlap of the laser beam and the telescope field of view starts beyond certain ranges, respectively r_0 and r_{fo}. For ranges $r < r_0$ the laser beam is completely out of the telescope field-of-view, i.e., there is "no overlap". For ranges $r_0 < r < r_{fo}$ the laser beam is only partially in the telescope field-of-view, i.e., there is a "partial overlap". The ranges $r > r_{fo}$ are where the laser beam is completely in the telescope field-of-view, i.e., there is the "full overlap".

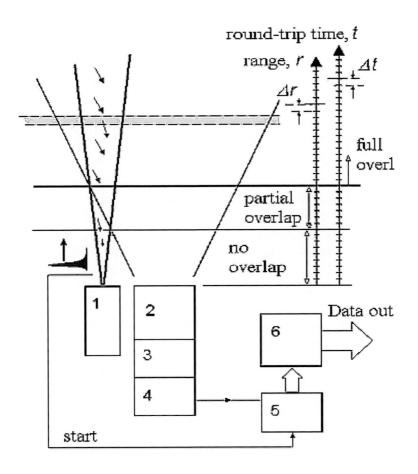

Figure 1. Illustration of the Atmospheric Lidar principle and set-up. Legend: 1. Laser; 2. Telescope; 3. Filter; 4. Detector; 5. Signal acquisition electronics; 6. Processing unit.

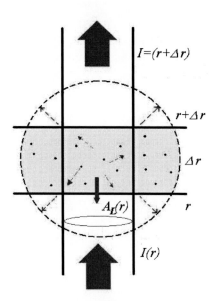

Figure 2. Illustration of the derivation of the lidar equation: scattering and attenuation of the laser beam in the atmosphere.

A dedicated optical component or subset is placed at the back-end of the telescope. Its function is to separate the optical radiation at the probing wavelength (or wavelengths, if more than one is used) and to cut the ambient optical background at the other wavelengths, and to separate the informative components of the backscatter light (if more than one) from each other. Depending on the lidar type, as well as if two or more wavelengths are used, this may be either just an interference filter or a more sophisticated combination of beamsplitters, filters, polarization analysers, spectrometers and interferometers.

After its separation, the collected backscattered light is directed to the optical detector. Photomultipliers are typically used in the ultraviolet and visible spectral ranges, and sometimes in the near infrared. Avalanche photodiodes are used in the near infrared range. The electrical signal from the detector (or the detectors, in case more than one light components are detected simultaneously) is digitised and registered by a signal detection and acquisition system.

After the start, in synchronisation with the transmitted laser pulse, the signal acquisition electronics acquires the electrical signal from the detector arriving at time t and in a time-interval Δt. This signal corresponds to light arriving from range $r = ct/2$, where c is the speed of light. This relation follows from the considerations that the laser pulse shall reach this volume and then the light backscattered from it shall arrive back to the lidar (round-trip time). Following the same considerations, the range-resolution corresponding to the acquisition time-bin Δt, is $\Delta r = c\Delta t/2$.

Two types of signal detection are used in the atmospheric lidars: photon-counting and analog [*Jenkins 1987*]. When the optical backscatter signal is with low power, each photon may produce a distinguishable electrical pulse and to be counted separately. This is referred to as photon-counting detection. When the backscattered light has sufficient power, the electrical pulses produced by the arriving photons completely overlap. In such a case the detector current is measured by an analog-to-digital converter. In both detection modes, the digitalised signal may be accumulated over a number of laser pulses (or sweeps). After that, the accumulated signal is sent to a computer for storage and further processing.

2.2. The Lidar Equation

Let us consider transmission of a single light pulse of rectangular shape, with energy E_L, duration τ_L and wavelength λ. At distance r, the laser beam is spread over an area $A_L(r)$, as illustrated in Figure 2. During the travel to r, the light intensity has decreased due to the extinction and absorption. The irradiance over the area $A_L(r)$ is expressed as

$$I_\lambda(r) = \frac{E_L TR_\lambda(r)}{\tau_L A_L(r)} \tag{1}$$

Here $TR_\lambda(r)$ is the transmission factor linked to the optical extinction $\alpha_\lambda(r)$ through the Beer-Lambert law [*Measures, 1984*]

$$TR_\lambda(r) = exp\left(-\int_0^r \alpha_\lambda(r')dr'\right) \tag{2}$$

Restricting to one type of scatterer with number density $n(r)$, and neglecting presently the absorption, $\alpha_\lambda(r)$ takes the form

$$\alpha_\lambda(r) = \alpha_\lambda^{sca}(r) = n(r)\int\frac{d\sigma_\lambda}{d\Omega}d\Omega \tag{3}$$

Here $\left(\dfrac{d\sigma_\lambda}{d\Omega}\right)$ is the differential scattering cross-section. The integral accounts for scattering over the unit sphere, relatively to a spherical coordinate system (θ,φ) with the polar axis in the direction of the incident beam, $d\Omega = \sin\theta\, d\theta\, d\varphi$. For a mixture of different scattering species we apply a superposition principle, and Equation (3) becomes

$$\alpha_\lambda(r) = \sum_i n_i^{scatt}(r)\int\frac{d\sigma_\lambda^i}{d\Omega}d\Omega \tag{4}$$

where $n_i^{scatt}(r)$ and $\dfrac{d\sigma_\lambda^i}{d\Omega}$ are respectively the number density and the differential cross-section of the ith scattering species. If λ lies in the absorption band of some molecular species, the absorption is accounted by adding the following additive term

$$\alpha_\lambda^{abs}(r) = n^{abs}(r)\sigma_\lambda^{abs} \tag{5}$$

In Equation (5) $n^{abs}(r)$ and σ_λ^{abs} are respectively the number density and the absorption cross-section of the species at wavelength λ. The volume backscatter coefficient is defined as

$$\beta_{\lambda,\lambda'}(r) = n^{scatt}(r)\left[\frac{d\sigma_{\lambda,\lambda'}}{d\Omega}\right]_{\theta=\pi} \tag{6}$$

In Equation (6) the term in the bracket is the differential scattering cross-section at wavelength λ' with incidence wavelength λ. If the backscatter process is elastic, then the incidence (laser) and scattered wavelengths are the same. If the scattering is Raman, then the backscatter wavelength is different from the incident, $\lambda' \neq \lambda$. As the pulse is transmitted at time $t = 0$, then at $t = r/c$ only the volume $V(r) = A_L c\tau_L / 2$ contributes to the optical power $P^b_{\lambda,\lambda'}(r)$, backscattered in the acceptance solid angle $\Delta\Omega(r)$ of the receiver telescope, when the latter is viewed from distance r

$$P^b_{\lambda,\lambda'}(r) = I_\lambda(r)\beta_{\lambda,\lambda'}(r)V(r)\Delta\Omega(r) \tag{7}$$

The acceptance solid angle $\Delta\Omega(r)$ is proportional to the telescope area A.

$$\Delta\Omega(r) = \frac{A}{r^2} \tag{8}$$

Travelling back to the lidar, the backscattered light is attenuated at wavelength λ'. The overlap between emitted laser beam and telescope field of view is represented by a function $O(r)$, where $O(r) = 0$ for $r < r_0$; $O(r) = 1$ for $r > r_{fo}$; $0 < O(r) < 1$ for $r_0 < r < r_{fo}$. The optical power received by the telescope is

$$P^{tel}_{\lambda,\lambda'}(r) = O(r)P^b_{\lambda,\lambda'}TR_{\lambda'}(r) \tag{9}$$

To obtain the detected electrical power $P_{\lambda,\lambda'}(r)$ we include in Equation (9) the optical efficiency of the transmitter and receiver K, the wavelength dependant efficiency of the detector $\eta_{\lambda'}$, as well as equations (1, 2, 7, 8).

$$P_{\lambda,\lambda'}(r) = E_L K\eta_{\lambda'}O(r)\frac{A}{r^2}\frac{c}{2}\beta_{\lambda,\lambda'}(r)TR_\lambda(r)TR_{\lambda'}(r) \tag{10}$$

During a single-pulse integration time Δt, this yields a detected energy

$$E_{\lambda,\lambda'}(r) = E_L K \eta_\lambda O(r) \frac{A}{r^2} \frac{c\Delta t}{2} \beta_{\lambda,\lambda'}(r) TR_\lambda(r) TR_{\lambda'}(r) \qquad (11)$$

When $\lambda = \lambda'$ (elastic backscattering), Equation (11) takes the form

$$E_\lambda(r) = E_L K \eta_\lambda O(r) \frac{A}{r^2} \frac{c\Delta t}{2} \beta_\lambda(r) exp\left(-2\int_0^r \alpha_\lambda(r')dr' \right) \qquad (12)$$

Equations (11) and (12) present the Lidar equation respectively for Raman and for elastic scattering.

The single signal acquisition bin Δt cannot be smaller than the temporal resolution of the overall measurement process. The lower limit for the range resolution is

$$\Delta r \geq \frac{c}{2}\left(\tau_L + \tau_D + \tau_P\right) \qquad (13)$$

Where τ_D, and τ_P are respectively the detection response time and the optical interaction process lifetime. For elastic and Raman scattering processes, τ_P is negligible.

In both extinction and volume backscatter coefficients in Equation (12) the contribution from molecules and from aerosol particles may be separated in the following way

$$\alpha_\lambda(r) = \alpha_\lambda^{mol}(r) + \alpha_\lambda^{aer}(r) \qquad (14)$$

$$\beta_\lambda(r) = \beta_\lambda^{mol}(r) + \beta_\lambda^{aer}(r) \qquad (15)$$

The molecular contributions may be evaluated independently from the density of the atmosphere and molecular constants as we will see further. The atmospheric density may be obtained either from atmospheric models or from pressure and temperature profiles measured by meteorological radiosondes.

Here we may define the "total backscatter ratio", referred also as "total scattering ratio"

$$scr_\lambda(r) = \frac{\beta_\lambda^{mol}(r) + \beta_\lambda^{aer}(r)}{\beta_\lambda^{mol}(r)} \qquad (16)$$

2.3. Overlap Function

The following specifications of the transmitter and receiver optics define the overlap function $O(r)$ and the ranges r_0, and r_{fo} : divergence of the transmitted beam, receiver field of view, angle between the axis of the transmitted beam and the optical axis of the receiver, distance between the beam center and transmitter axis at the lidar optical assembly. Examples for such function for two realistic configurations of lidar optics are shown in Figure 3, calculated in accordance to [*Measures, 1984*].

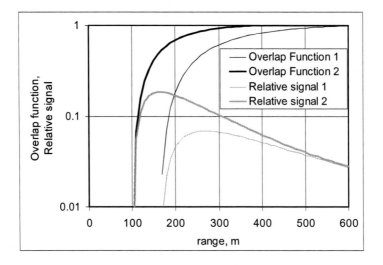

Figure 3. Overlap function (in absolute units) and the relative received backscatter signal in arbitrary units) for two examples of lidar optical set-up. The examples 1 and 2 are calculated with the same telescope diameter and field of view, and for the same laser beam divergence and power. The difference is in the distance between the axis of the telescope and the laser beam. For example 2 this distance is larger.

2.4. Noise Components in the Detected Lidar Signal

In the output signal from the detector, the noise components shall be added to the signal determined by equations (11, 12). Here follow a brief overview of the principle noise components in the backscatter lidar signal [*Candy 1985, Jenkins 1987, Overbeck et al 1995*].

2.4.1. Quantum or Shot Noise

We can consider the photons arrivals as independent random events with distribution represented by a Poisson statistic. In photon-counting detection, the photoelectrons are counted with associated values "1" (arrival) or "0" (no arrival), with a probability η_λ (here and further the quantum efficiency). Equation (12) may be modified to expresse the mean number of detected photoelectrons in interval Δt

$$N_s(r) = \frac{\eta_\lambda}{h\upsilon} E_L K \eta_\lambda O(r) \frac{A}{r^2} \frac{c\Delta t}{2} \beta_\lambda(r) exp\left(-2\int_0^r \alpha_\lambda(r')dr'\right) \qquad (17)$$

Here υ is the optical frequency and h is the Planck's constant ($h\upsilon$ is the energy of the incoming photon). The photon detection is a random event, where the individual realisations fluctuate around the mean value. The noise associated with these fluctuations is called quantum or photon noise. In Poisson statistic the variance is equal the mean value of the number, i.e.

$$\sigma^2_{N_s} = \overline{N}_s \tag{18}$$

In analog detection mode, the signal determined by Equation (12) give rise to an instantaneous value of the current at the output of the detector (photomultiplier or avalanche photodiode)

$$i_s(r) = G\frac{\eta e}{h\upsilon} BE_L K\eta_\lambda O(r)\frac{A}{r^2}\frac{c\Delta t}{2}\beta_\lambda(r)exp\left(-2\int_0^r \alpha_\lambda(r')dr'\right) \tag{19}$$

Here G is the internal gain of the detector, $B = 1/\Delta t$ is the electrical bandwidth. The noise associated with the fluctuations of the instantaneous current around its mean value is called shot noise with a variance

$$\sigma^2_{i_s} = 2eBG\mu i_s \tag{20}$$

In Equation (20) μ is the excess noise factor, typically $\mu > 1$.

2.4.2. Optical Background and Dark Noise

Any light source other than the backscattered light contributes to the input to detector and leads to an additional signal. Such sources are either natural, as the sun or moon light scattered from sky and surfaces, or artificial. For the illuminated sky, the power of the received background light is

$$b(t) = K_r S_b(\lambda)\Delta\lambda\Omega_r A \tag{21}$$

In Equation (21) K_r is the optical efficiency of the receiver, S_b (in [$Wm^{-2}sr^{-1}\mu m^{-1}$]) is the sky spectral radiance at the instrument location and instant of measurement t, $\Delta\lambda$ (in [μm]) is the optical bandwidth of the receiver, Ω_r in steradian, the receiver field of view (not to be confused with Equation(8)) and A, as before, is the receiver aperture area. The receiver field of view is linked to the full Field-of-View angle ψ of the receiver by

$$\Omega_r = \pi\left(\frac{\psi}{2}\right)^2 \tag{22}$$

The background photons number N_b and the background current i_b are expressed as

$$N_b(t,\tau) = b(t)\Delta t\frac{\eta}{h\upsilon} \tag{23}$$

$$i_b(t) = b(t)G\frac{\eta e}{h\upsilon} \tag{24}$$

The natural background can be considered as constant during the acquisition. Thus, it appears as an offset to be subtracted from the measured signal. Its effect is to increase the fluctuations of the mean photon number or mean electrical signal produced by the detector, following considerations as for equations (19, 20)

$$\sigma^2_{N_b} = N_b \tag{25}$$

$$\sigma^2_{i_b} = 2eBG\mu i_b \tag{26}$$

In absence of incoming radiation, the photo-detectors produce currents due to the thermal emission or thermal current, referred as dark current. In photon-counting, the detector is characterized by a temperature-dependent dark count rate d_l, such that the number of dark counts in the single-pulse detection period is

$$N_d(t,\tau) = d_l\Delta t \tag{27}$$

In analog detection mode, this gives rise to the dark-current i_d.

The effect of the optical background and the dark current (counts) is in the increase of the fluctuations. The variances of these increases, respectively for photon counting and analog detection are

$$\sigma^2_{N_d} = N_d \tag{28}$$

$$\sigma^2_{i_d} = 2eBG\mu i_d \tag{29}$$

2.4.3. Thermal (Johnson) Noise

In analog detection the PMT output current is passing through an electronic chain and the output signal is measured as a variable voltage. Thermal fluctuations of the charge carriers within the electronic chain lead to an additional fluctuation in the measured signal. The variance of this current is again a function of the temperature

$$\sigma^2_{th} = \frac{4k_B T R_{ch} B_e}{R_{ch}} \tag{30}$$

Here k_B is the Boltzmann constant, $T_{R_{ch}}$ the absolute temperature of the resistance, B_e is the electrical bandwidth and R_{ch} the equivalent resistance of the electronics chain (amplifier, etc.).

2.4.4. Signal-Induced Noise and Detector Linearity

The signal-induced noise is an echo of the previously arrived light. In the photomultipliers this is also known as an after-pulsing effect and it contributes to a systematic error in the lidar signal measurement [Coates 1973 and 1973a, Antonioly and Benetti 1983]. As the signal from ranges close to the lidar is much stronger than the signal from longer ranges, the after-pulsing adds a "false" signal to the signal received from the longer ranges.

As we see from equations (10-12) and Figure 3, due to the r^{-2} dependence the detected signal decreases sharply with the range. The terms $T_\lambda(r), T_{\lambda'}(r)$ contribute to the decrease of the signal from further ranges too, as well as the altitude dependence of $\beta_{\lambda,\lambda'}(r)$ if probing upward (see further in Chapter 3). I.e., to have proportionality between the recorded backscattered signal and the atmospheric parameters, it is necessary that the optical detector gives a linear response to the optical signal for all ranges of interest. This may be achieved by convenient technical solutions or by appropriate correction schemes [Donovan et al. 1993, Bristow et al. 1995].

2.4.5. Signal-to-Noise Ratio

The signal-to-noise ratio (SNR) is a measure of the quality of the atmospheric backscatter signal extracted from the measured (noisy) signal. It is defined as

$$SNR(r) = \frac{E(r)}{\sqrt{\sum_k \sigma_k^2(r)}} \tag{31}$$

In Equation (31) $E(r)$ is the atmospheric backscattered signal and the $\sigma_k(r)$ are the standard deviations of the statistically independent noise sources, i.e., the contributions in 2.4.1-2.4.3, where effects noted in 2.4.4 are supposed to be corrected.

For the photon-counting case, this takes the following form

$$SNR(r) = \frac{N_s(r)}{\sqrt{N_s(r) + N_b(r) + N_d(r)}} \tag{32}$$

In the analog case, we have to add also the thermal (Johnson) noise

$$SNR(r) = \frac{i_s(r)}{\sqrt{2eBG\mu(i_s(r) + i_b(r) + i_d(r)) + \dfrac{4kT_{R_{ch}}B}{R_{ch}}}} \tag{33}$$

3. ATMOSPHERE

3.1. Atmospheric Structure and Major Constituents

The atmosphere is the mixture of gases enveloping our planet [*Salby 1996*]. Table 1 presents the relative concentration of the atmospheric gases. Nitrogen and oxygen constitute more than 99 % in volume of the permanent gases with a nearly constant volume mixing ratio. The vertical density profiles of the important atmospheric constituents are described in the model US-76 [*US Standard Atmosphere 1976, Anderson et al. 1986*]. Concerning the interpretation of the lidar measurements, the molecular atmosphere is well described by the ideal gas law in terms of altitude z

$$p(z) = n_{molec}(z)k_B T(z) \tag{34}$$

In Equation (34) $p(z)$ is the pressure, $n_{molec}(z)$ is the molecular number density, $T(z)$ is the absolute temperature. The hydrostatic equation is approximating the vertical distribution of the atmospheric pressure

$$p(z) = p_0 \exp\left(-\int_0^z \frac{g(z')}{RT} dz' \right) \tag{35}$$

or in differential form

$$dp(z) = -\rho(z)g(z)dz \tag{36}$$

$\rho(z) = n_{molec}(z)M_{air}$, p_0 is the surface pressure, R is the ideal gas constant, M_{air} is the average molecular mass of air, g the acceleration of gravity. The operational device to obtain $n_{molec}(z)$ is the meteorological radiosonde, carrying sensors for $T(z)$ and $p(z)$. Equations (35-36) may be applied to about 100km altitude. In the lower atmosphere, the pressure decreases by a factor 2 pro 5 km altitude increase. However, such altitude distribution is valid for nitrogen and the oxygen, but not for many minor components.

The vertical temperature profile of the atmosphere is determined by the balance between the incident solar radiation and the emission of radiation from the Earth. Figure 4 presents the temperature profile of the standard atmospheric model US-76 to 100 km above sea level (asl). The sign of the temperature lapse rate defines the different atmospheric parts. In the troposphere, the lowest part of the atmosphere, the mean temperature lapse rate is about 6.5°C/km. Nearly all water vapour, clouds and precipitation are confined in the troposphere. Its upper boundary is the tropopause, a temperature minimum, appearing at an altitude of about 9-10 km in the Polar Regions to 15-17 km in tropics. The stratosphere contains the main part of the atmospheric ozone, preventing the ultraviolet (UV) light to reach the Earth's surface. A balance between radiative cooling from CO_2 and heating by O_3 mainly controls the temperature distribution in the upper stratosphere and lower mesosphere. The temperature profile is one important target for atmospheric probing.

Table 1. Composition of the atmosphere, following US-76 and [Anderson 1986]([a] is for concentration near the Earth's surface)

Permanent Constituent	Part per Volume [%]	Variable Constituent	Part per Volume [%]
Nitrogen (N_2)	78.084	Carbon monoxide (CO) [a]	0.19×10^{-4}
Oxygen (O_2)	20.946	Xenon (Xe)	0.089×10^{-4}
Argon (Ar)	0.934	Water-vapour (H_2O)	0-4
Carbon dioxide (CO_2)	0.033	Ozone (O_3)	$0-12 \times 10^{-4}$
Neon (Ne)	18.18×10^{-4}	Sulphur dioxide (SO_2) [a]	0.001×10^{-4}
Helium (He)	5.24×10^{-4}	Nitrogen dioxide (NO_2) [a]	$0.001 \times 10^{-4)}$
Methane (CH_4)	1.5×10^{-4}	Ammonia (NH_3) [a]	0.004×10^{-4}
Krypton (Kr)	1.14×10^{-4}	Nitric oxide (NO) [a]	0.0005×10^{-4}
Hydrogen (H_2)	0.5×10^{-4}	Hydrogen sulphide (H_2S) [a]	0.00005×10^{-4}
Nitrous oxide (N_2O) [a]	0.27×10^{-4}	Nitric acid vapour (HNO_3)	Trace

The lowermost part of the troposphere is in contact with the Earth's surface and is named Planetary Boundary Layer (PBL) [*Stull, 1988*]. The surface induces a drag on the geostrophic wind, causing turbulences. The surface also creates the fluxes: evaporation (water vapour), reflection and thermal emission (energy), erosion (aerosol) and human activity (pollution). The PBL tends to accumulate aerosol, water vapour and pollution gases. Its structure depends on the time of the day and on the surface (day, night, flat land, mountain valley, sea surface, etc.). During the day, the combined action of heat, evaporation and turbulence leads to uplift and mixing of the air, containing water vapour, gases and aerosol particles emitted from the surface. The height at which this mixing takes place is named mixing layer height (MLH). The human life and activity takes place predominantly in the PBL. That is why this layer is important target for atmospheric monitoring.

3.2. Important Minor Gas Constituents

The important minor constituents in the atmosphere are both from natural origin and from human activity [*Schoulepnikoff et al. 1998*]. Also, some of the important minor constituents from natural origin are strongly affected by human activity. The altitude distribution of the number density of most of the minor gaseous components does not follow equations (35, 36). The reason for this is that the specific gas may be produced at a certain altitude and then undergo transformation at another altitude. The role of some minor constituents in the atmospheric processes and on life is critical; this explains why their monitoring is required.

Water vapour is a principal minor constituent. Its concentration and effect are such that it likely shall not be referred to as minor. It is the basic transporter of energy by evaporation, condensation and precipitation, and it has a deterministic effect on the climate and life on Earth. Due to its absorption spectrum, it is also important in the radiative absorption and emission processes. The vertical distribution of the water vapour number density does not follow the low of exponential decrease, since the evaporation takes place at surface level; the transport and condensation occur at altitudes with follow-on precipitation.

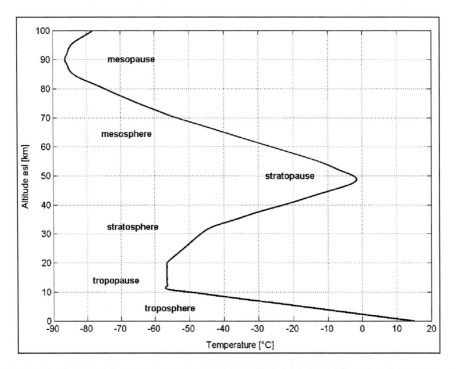

Figure 4. Vertical profile of the atmospheric temperature, based on the standard US-76 model.

Although CO_2 is listed in Table 1 as a permanent constituent, its concentration varies as a result of the combustion of fossil fuels, absorption and release by the ocean and photosynthesis. The recent trend of its concentration is an increase of approximately 1.5 ppmv/year. This gas is transparent to visible and near infrared (IR) solar radiation, but absorbs in the 12-18 µm range (terrestrial radiation). Due to this, it plays a major role in the climate changes, known as the "greenhouse" effect. Another important "greenhouse" gas is methane. Large natural sources of methane are the marshes. A large contribution comes also from agriculture, natural gas storage and transportation, etc.

Ozone is another important minor constituent. Its concentration also does not follow the exponential decrease, since it is produced in the stratosphere by solar radiation. The ozone in the stratosphere plays a vital role in absorbing harmful ultraviolet solar radiation and in this way creating conditions for life on Earth. Its concentration is threatened by the release of gases known as chlorofluorocarbons (CFC's) into the atmosphere. These compounds, in the specific conditions of the winter Polar stratosphere, cause the phenomenon known as the "ozone hole", observed since the 1980s.

Ozone is also produced in the PBL and the troposphere, by photochemical reactions involving anthropogenic pollution gases as nitrogen oxides, CO, hydrocarbons, etc. Due to this, the ozone concentration depends on the pollution source and the meteorological conditions, and is variable in space and time. As it is chemically very active (oxidant), it has a harmful effect on life.

A number of small gaseous components in the atmosphere result from human activity: SO_2, CO, NO, NO_2, etc. Their importance is due to the harmful effect that they have on life. This effect is increased by the fact that their emission and transport take place close to populated areas [*Schoulepnikoff et al. 1998*].

3.3. Atmospheric Aerosol

Aerosols are tiny particles, suspended in the air and carried with its movement. Aerosol particles may be solid or liquid. They range in size from 0.01μm to several tens of μm. A survey of the types of aerosol, its global climatology and effects may be found in [*d'Almeida et al. 1991, Godish 1991*].

Aerosols are considered a major uncertainty source in the climate models. On one hand, through their scattering, they contribute to the total solar radiation backscattered to space. Aerosols also affect the properties and the amount of the clouds, acting as condensation nuclei for cloud formation. By this they may affect cloud occurrence and thickness, rainfalls, and once again the amount of sunlight that is reflected back to space. Such effects may play the role of negative feedback in the processes referred to as global warming. On the other hand, through their own absorption, aerosols contribute to atmospheric warming.

Near the surface aerosols may be also a harmful pollution agent.

In the stratosphere there are two major types of aerosol. One is the background stratospheric aerosol [*Turco et al. 1982*]. Those are small droplets of sulphuric acid mixed with water. They originate from gases released by volcano eruptions and occur in the altitudes between 12km and 30km. Another type, named polar stratospheric clouds (PSC), is formed above the Polar regions at altitudes of approximately 17km-27km. PSC particles consist of ice-crystals composed either of water or from a mixture of nitric acid and water. They play a critical role in the triggering the chemical processes leading to the destruction of the stratospheric ozone layer.

The principal aerosol types in the troposphere are classified as follows: *Soot aerosol* consists of black carbon, originating from fires and combustion. *Sea-salt aerosol* originates from sea-spray. *Mineral aerosol* is dust produced by wind erosion in the deserts areas, consisting of clay, silicates and other mineral compounds. When the desert dust is transported at long ranges by the wind, the mineral aerosol absorbs water and it is referred to as *mineral transported aerosol.*

Some aerosols are generated by human activities, such as burning of fossil fuels, traffic, industry, mining, etc. This brings about 10% of the total amount of aerosols in the troposphere, mainly concentrated in the Northern Hemisphere. Aerosol types are also defined in accordance with the anthropogenic sources and the place of observation: *continental aerosol* (respectively *continental clean* or *rural, continental average* and *continental polluted*), *marine clean* and *marine polluted aerosol, urban* or *industrial aerosol, Arctic aerosol* or *Arctic haze.*

Important characteristics of the aerosol related to the atmospheric lidars are their size distribution and number density. It is accepted that the number of the aerosol particles $n_i(\tilde{r})$ from the same single source i with the size (effective radius) $\tilde{r}$, follows the so-called lognormal distribution [*Ansmann and Müller 2005*]

$$\frac{dn_i(\tilde{r})}{d(\log \tilde{r})} = \frac{n_{i,t}}{(2\pi)^{1/2} \log \sigma_i} exp\left[-\frac{(\log \tilde{r} - \log \tilde{r}_{i,mod})^2}{2(\log \sigma_i)^2} \right] \qquad (37)$$

In Equation (37) $n_{i,t}$ is the total particle number density, σ_i is the geometrical standard deviation; and $\tilde{r}_{i,mod}$ is the median effective radius of the particle ensemble. When the aerosol is a mixture of several independent sources, the resulting distribution is a sum of the individual distributions

$$\frac{dn(\tilde{r})}{d(\log \tilde{r})} = \sum_i \frac{dn_i(\tilde{r})}{d(\log \tilde{r})} \qquad (38)$$

The mean size of the continental and urban aerosol is in the range of a few hundreds nm, while their number density is in the order of 10^4-10^5 particles per cm^3. The size of the desert dust particle is in the order of one μm with typical number density of 2000-3000 particles per cm^3 [*Ansmann and Müller 2005*]. The size of the background stratospheric aerosol is a few hundreds of nm with a number density of a few particles per cm^3 [*Turco et al. 1982*]. The importance of the aerosol for the atmospheric processes and for life motivates the efforts for their monitoring.

4. OVERVIEW OF THE SCATTERING AND ABSORPTION OF THE LASER BEAM IN THE ATMOSPHERE

In the atmospheric lidar probing, we take into account linear processes occurring in the propagation of optical radiation in the atmosphere. These processes involve two photons: one incident and one scattered. In the elastic molecular (Rayleigh) scattering, and in the elastic aerosol (Mie) scattering the frequency of the scattered photon is equal to the frequency of the incident one. In the Raman scattering the frequency of the scattered photon is shifted from the frequency of the incident one.

In the elastic scattering the properties of the light scattering depend on the size parameter defined as $x = 2\pi \tilde{r} / \lambda$, where $\tilde{r}$ is the mean radius of the particle and λ is the wavelength of the incident light, on their shape and on their refractive index (i.e., composition). For $x \ll 0.1$, the scattering is "Rayleigh" or "molecular". For larger x, that is the case of the atmospheric aerosol, the scattering type is referred as "Mie".

4.1. Rayleigh Scattering

In this brief overview of the Rayleigh scattering we consider a propagation of a plane wave with wavelength λ in coordinate system xyz. The propagation is in direction z and the polarisation is in direction y. The scattered wave is described in polar coordinates (θ, φ), where θ is the angle in plane zy and φ in xy. The molecular (Rayleigh) scattering differential cross-section is given as [*Penndorf 1957*]

$$\frac{d\sigma_R}{d\Omega} = \frac{9\pi^2(n_{air}^2-1)^2}{N_{air}^2\lambda^4(n_{air}^2+2)^2}\frac{6+3\rho}{6-7\rho}(cos^2\varphi\,cos^2\theta + sin^2\varphi) \tag{39}$$

In Equation (39) N_{air} is the number density of air molecules, n_{air} is the air refractive index [*Edlen 1966*] and ρ is the depolarisation ratio. The term $(6+3\rho)/(6-7\rho)$ accounts for anisotropy in molecular polarisability (King factor)

$$F(air) = \frac{6+3\rho}{6-7\rho} \tag{40}$$

The accepted value for dry air is $F=1.0481$ or $\rho=0.0279$) [*Young 1981*]. The scattering cross-section is determined as

$$\sigma_R = \int\frac{d\sigma_R}{d\Omega}d\Omega = \frac{24\pi^3(n_{air}^2-1)^2}{N_{air}^2\lambda^4(n_{air}^2+2)^2}F \tag{41}$$

The extinction coefficients is then determined as

$$\alpha_R = N_{air}\int\frac{d\sigma_R}{d\Omega}d\Omega \tag{42}$$

The backscattering (differential) cross-section is

$$\beta_R = N_{air}\sigma_R^\pi = N_{air}\left[\frac{d\sigma_R}{d\Omega}\right]_{\theta=\pi} = \frac{9\pi^2(n_{air}^2-1)^2}{N_{air}^2\lambda^4(n_{air}^2+2)^2}F \tag{43}$$

The ratio of molecular extinction-to-backscatter (or molecular "lidar ratio") is $lr^{mol} = \frac{8\pi}{3}$ sr.

The λ^{-4} dependency in Equation (41, 43) implicates that for probing the molecular component of the atmosphere, it is advantageous to use lasers in the UV. Discrepancies from this dependence emerge from the wavelength dependence of the air refractive index [*Eltermann, 1968*]. Table 2 shows Rayleigh scattering cross-section of dry air for some laser wavelengths and for CO_2 concentrations 365 ppmv.

Table 2. Rayleigh cross-section of dry air with 365 ppmv CO_2 for the wavelengths of lasers used in atmospheric lidars

Wavelength	266 nm	355 nm	532 nm	1064 nm
Cross-section	9.5598236 $\times 10^{-26}$ cm^2	2.7588752 $\times 10^{-26}$ cm^2	0.51673042 $\times 10^{-26}$ cm^2	0.031269537 $\times 10^{-26}$ cm^2

Figure 5 presents the angular distribution of the scattered intensity in function of the scattering angle θ for three values of φ ($\varphi = 0$ ('parallel'), $\varphi = \pi/2$ ('perpendicular'), $\varphi = \pi/4$ ('non polarised')). The term 'non polarised' is used here, since the angular distribution (versus θ) of the scattered intensity measured in the plane $\varphi = \pi/4$ is the same as the one from an incident non-polarised light (for which there is no φ dependency). In Figure 5, the radial scale gives the angular dependence of the volume scattering coefficient. Lets notice that the scattering process is with symmetry between forward and backward directions. In the plane parallel to the incident beam polarization, the scattered intensity at 90° vanishes.

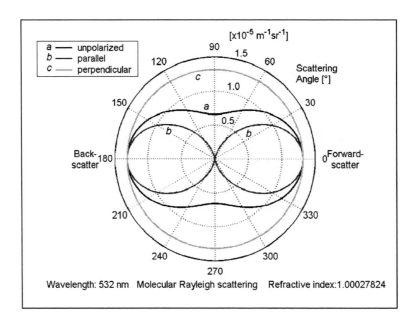

Figure 5. Angular distribution of the volume scattering coefficient β_R for a linearly polarized beam at 532nm, in dry air at 288.15K, 1013.25hPa. [Frioud 2003].

4.2. Mie Scattering

Any analytical description of light scattering from aerosol particles is beyond the scope of the present review. Here we may only illustrate some of its features relevant to the atmospheric lidars following [*Bohren and Huffman 1983*]. An overview on this subject may be also found in [*Measures 1984*].

We will consider that the particle number density is such that light scattered by one particle is received without being scattered once more from another particle, i.e. single-scattering approximation. Thus the scattering on each particle can be considered as independent from the other particles. We also restrict our example to monodispersion of homogeneous spherical particles. The scattering of such particles is described analytically by the Mie theory. Inputs in such considerations are the size parameter x and the relative refractive index m of the particle, practically equal to refractive index in vacuum for aerosol particle in the atmosphere.

The approach in [*Bohren and Huffman 1983*] defines corresponding efficiencies Q, for extinction, scattering and absorption, as follows:

$$Q^{ext} = \frac{\sigma^{ext}}{\pi \tilde{r}^2}; \ Q^{sca} = \frac{\sigma^{sca}}{\pi \tilde{r}^2}; \ Q^{abst} = \frac{\sigma^{abs}}{\pi \tilde{r}^2} \tag{44}$$

Here σ^{ext}, σ^{sca} and σ^{abs} are the extinction, scattering and absorption cross-sections, where $\sigma^{ext} = \sigma^{sca} + \sigma^{abs}$; $\tilde{r}$ is the particle radius, respectively $\pi \tilde{r}^2$ is the area of the particle cross-section projected onto a plane perpendicular to the propagating beam. In such approximation the (differential) backscatter efficiency is

$$Q^{back} = \frac{\sigma^{back}}{\pi \tilde{r}^2} = \frac{4}{\tilde{r}^2} \left[\frac{d\sigma^{sca}}{d\Omega} \right]_{\theta = \pi} \tag{45}$$

where the corresponding backscatter cross-section is

$$\sigma^{back} = 4\pi \left[\frac{d\sigma^{sca}}{d\Omega} \right]_{\theta = \pi} \tag{46}$$

The volume backscattering coefficient is then

$$\beta^{aer} = N^{aer} \frac{Q^{back} \pi \tilde{r}^2}{4\pi} \tag{47}$$

where N^{aer} is the particle number density. The particle extinction-to-backscatter ratio or "lidar ratio" is

$$lr^{aer} = \frac{\alpha^{aer}}{\beta^{aer}} = 4\pi \frac{Q^{ext}}{Q^{back}} \tag{48}$$

In the general case when the aerosol particles are not monodispersion, then Equation (47) shall take into account the particle size distribution, i.e., equations (37, 38).

Figure 6 presents example for extinction and backscatter efficiencies while Figure 7 shows lr^{aer} versus the size parameter, following the approach in [*Bohren and Huffman 1983*]. Figure 8 shows an example for the angular distribution of the volume backscatter coefficient for $x = 0.945$. The difference with the Rayleigh case (Figure 5) is the asymmetry between forward and backward scattering. This forward scattering predominance becomes extremely strong when $x \gg 1$. With increasing size parameter there is an increasing angular mode structure (succession of peaks and holes at large scattering angles). This may be seen in Figure 9, which presents the scattering pattern for for $x = 7.086$.

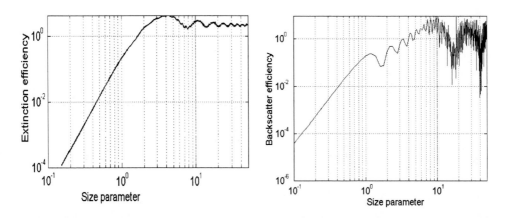

Figure 6. Example for aerosol extinction and backscatter efficiency as function on the size parameter. Mie-calculations with particle parameters: sphere, refractive index 1.5. [Frioud 2003]. Left: extinction efficiency; Right: Backscatter efficiency.

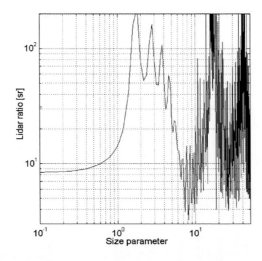

Figure 7. Example for aerosol lidar ratio as function on the size parameter. Mie calculations with particle parameters: sphere, refractive index 1.5 [Frioud 2003].

The presentaed examples are valid for ensembles of particles with the same size and refractive index. In reality, the aerosol in the atmosphere is a mixture of particles having different sizes and shapes, and composition (i.e., refractive index values), making the evaluation of the scattering parameters rather complicated. One example, for evaluation of lr^{aer} is reported in [*Ackermann, 1998*], based on Mie calculations for mixture of continental type aerosol.

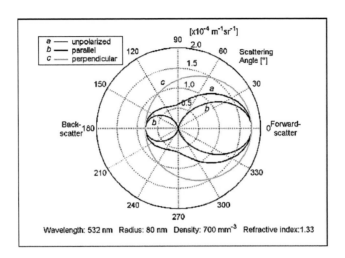

Figure 8. Angular distribution of the volume scattering coefficient for a linearly polarized beam at 532 nm, for a monodispersion of particles, with radius 80 nm, density 700 mm^{-3} and refractive index 1.33 (Mie); size parameter $x = 0.945$ [Frioud 2003].

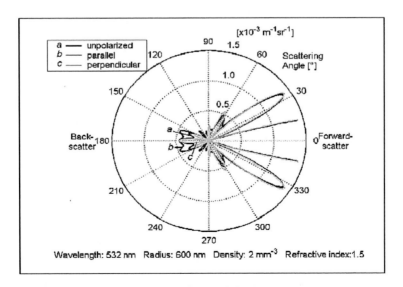

Figure 9. Angular distribution of the volume scattering coefficient for a linearly polarized beam at 532 nm, for a monodispersion of particles of radius 600 nm, density 2 mm-3 and refractive index 1.5 (Mie calculation); size parameter $x = 7.086$ [Frioud 2003]. The forward lobe extends to 8x10-3 m-1sr-1.

4.3. Spectral Broadening of the Rayleigh Scattering Line

The molecules in the atmosphere are in thermal motion with randomly distributed velocities. The velocities follow the distribution of Maxwell with width of approximately 200-300m/s. Assuming a δ-function-like laser line with optical frequency ν_L, this distribution leads to a broadening of the Rayleigh scattered line. The intensity distribution of the scattered line versus the frequency $I_s(\nu)$ is expressed as follow

$$I_s(\nu) = \sqrt{\frac{M_{air}\nu_L^2}{2\pi k_B T}}\ \exp\left[-\frac{M_{air}c^2(\nu-\nu_L)^2}{2k_B T\nu_L^2}\right] \tag{49}$$

Figure 10 presents the Doppler broadened Rayleigh line for air at temperature 293K.

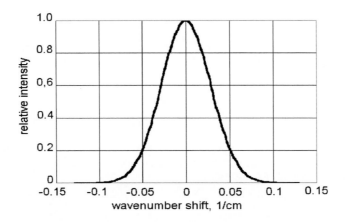

Figure 10. The spectral profile of the Rayleigh line of air at temperature 290K.

4.4. Raman Scattering

The Rayleigh and Mie scattering are elastic, i.e., the wavelength (further we will use frequency or wavenumber) of the scattered photons is the same as that of the incident one. The Raman scattering is inelastic, i.e., the scattered photon has frequency ν_{Ram} shifted from the laser frequency ν_L as $\nu_{Ram} = \nu_L \pm \Delta E_{Ram}/hc$, where ΔE_{Ram} is the energy difference between the molecular energy levels involved in the scattering. If the shift is with sign "-" the Raman scattering components are of "Stokes type", respectively when the shift is with sign "+" the Raman scattering components are of "anti-Stokes type" (illustrated in Figure 11). For each molecule the Raman shift is also specific, determined from the molecules energy levels and symmetry. In this way the Raman spectrum is a kind of a "finger print" for the respective molecule.

The Raman lines used in the atmospheric lidars are from the most abundant atmospheric constituents: nitrogen, oxygen and water vapour. The nitrogen and oxygen molecules are diatomic. Such molecules may both rotate and vibrate with respect to their center of mass.

The transitions between the vibrational levels with quantum number υ and between rotational levels with rotational quantum number J follow the selection rules

$$\Delta\upsilon = 0,\pm1 \qquad\qquad (50)$$

$$\Delta J = 0,\pm2$$

The energy of the vibrational level is

$$E_{vib,\upsilon} = hc\,v_{vib}(\upsilon+1/2) \qquad\qquad (51)$$

where v_{vib} is the vibrational wavenumber. The energy of the rotational level is defined as

$$E_{rot,\upsilon,J} = hc\left[B_\upsilon(J(J+1)-D_\upsilon J^2(J+1)^2\right] \qquad\qquad (52)$$

In Equation (52) B_υ and D_υ are respectively the rotational and the centrifugal distortion constant of the specific molecule for vibrational quantum number υ. In the terms of the above equations the Rayleigh scattering corresponds to $\Delta\upsilon = 0$ and $\Delta J = 0$. The lines with $\Delta\upsilon = 1$ form Stokes vibrational-rotational band while the lines with $\Delta\upsilon = -1$ form the anti-Stokes vibrational rotational band. For $\Delta\upsilon = \pm1$, the lines with $\Delta J = 0$ form the Q-branch of the vibrational rotational band, the lines with $\Delta J = +2$ form the S branch and with $\Delta J = -2$, respectively O-branch. The case $\Delta\upsilon = 0$ and $\Delta J \neq 0$ is the pure Rotational Raman scattering, with $\Delta J = +2$ for the Stokes branch and $\Delta J = -2$ for the anti-Stokes branch. Equation (52) also demonstrates that around each vibrational Raman line there is a number of lines determined from the rotation of the molecule, referred as rotational-vibrational spectrum.

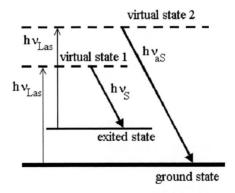

Figure 11. Illustration of the Raman Stokes and Raman anti-Stokes scattering; the subscripts "Las", "S" and "aS" are respectively for the *Laser, Stokes* and *anti-Stokes* photons.

At the typical atmospheric temperatures, the population of the state $\Delta\upsilon = 1$ is negligible compared to the ground state. Respectively, the intensity of the anti-Stokes scattering will be negligible compared to the Stokes one. That is why in the atmospheric Raman lidar using vibrational lines and bands, only Stokes lines are used.

The calculation of the vibrational Raman backscatter cross-section of the Stokes band for nitrogen and oxygen in non-resonant conditions may refers back to [*Placzek 1934, Long 2002*] and to the summary presented in [*Wandinger 2005*]

$$\left[\frac{d\sigma_{vib}(v)}{d\Omega}\right]_{\theta=\pi} = \frac{k_v(v_L - v_{vib})^4 b_v^2}{[1 - exp(-hcv_{vib}/k_BT)]}\left(a'^2 + \frac{7\gamma'^2}{45}\right) \tag{53}$$

Here $k_v = \dfrac{\pi^2}{\varepsilon_0{}^2}$, where ε_0 is the permittivity of the vacuum. The parameters a' and γ' are respectively the derivatives of the mean polarisability (a) and the anisotropy of the polarisability (γ) with respect to the normal vibrational coordinate of the molecule in equilibrium position. The value b_v^2 is the square of the zero-point amplitude of the vibrational mode and is defined as

$$b_v^2 = \frac{h}{8\pi^2 c v_{vib}} \tag{54}$$

The atmospheric Raman lidars use v_{vib} for $\upsilon = 1$ for nitrogen and oxygen. The frequency shifts are respectively for nitrogen 2330.1cm^{-1} and for oxygen 1556.4cm^{-1}. It shall be noted that the rotational structure appearing around the vibrational lines has width of a few nm. As we see, the vibrational Raman cross-section (but not its rotational structure), for temperatures typical for the atmosphere may be considered a molecular constant.

In the rotational Raman scattering the population of the rotational levels is determined by both the thermal distribution and the statistical weight (frequency of occurrence) [*Herzberg 1950*]. Each molecular state with angular momentum J consists of $2J + 1$ levels coinciding in absence of external field. Thus the statistical weight of a rotational state increases with the number J and the population is as follows

$$N_{rot}(J) = (2J + 1)exp(-hcv_{rot}/k_BT) \tag{55}$$

where v_{rot} is the vibrational wavenumber.

This increase, combined with the lower values for v_{rot}, provides significant values for the pure rotational Raman cross-sections, comparable for Stokes and anti-Stokes lines. Following [*Herzberg 1950*], the scattering cross-section for rotational quantum number J may be expressed as

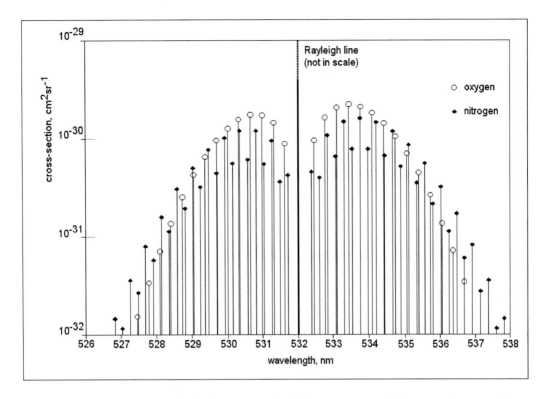

Figure 12. Molecular cross-section in the pure rotational Raman spectra of nitrogen and oxygen. The incident wavelength is 532nm.

$$\left[\frac{d\sigma_{rot}(J)}{d\Omega}\right]_{\theta=\pi} = [\nu_L - \nu_{rot}(J)]^4 g_I \frac{B}{k_B T} (2J+1)s(J)\exp[-BJ(J+1)/k_B T] \quad (56)$$

Where B is the rotational constant for $\upsilon = 0$, g_I is the statistical weight due to the nucleus spin. The product of the function $s(J)$ and the statistical weight is expressed for the Stokes lines as

$$(2J+1)s(J) = \frac{(J+1)(J+2)}{(2J+3)} \quad (57)$$

and for the anti-Stokes lines as

$$(2J+1)s(J) = \frac{J(J-1)}{(2J-1)} \quad (58)$$

An interesting feature of the pure rotational Raman lines of nitrogen and oxygen is that the temperature derivative of the cross-section is negative for smaller J and positive for larger J, for both Stokes and anti-Stokes parts. This is the base for the rotational Raman lidar method for temperature measurements. Figure 12 presents the pure rotational Raman scattering spectra for nitrogen and oxygen for incident wavelength 532nm.

Water vapour is another species of high interest in atmospheric studies. This is a molecule from the type of nonlinear asymmetric top molecule. There are three normal vibration modes, also coupled with rotation, which make the water Raman spectrum quite complicated. Here we will note only that the strongest line, the one having practical application in the atmospheric Raman lidars is the symmetric elongation mode, referred as $\upsilon 1$. The vibrational wave-number shift of this mode is 3651-3652nm. The line $\upsilon 1$ also demonstrates well developed rotational structure. The spectrum of this molecule is discussed in detail in [*Avila et al. 1999*].

Values of the vibrational Raman cross-sections are determined in a number of studies [*Bischel and Black 1983, Faris and Copeland 1997*]. Calculated spectra of nitrogen and oxygen are presented in [*Wandinger 2005*]. The scattering cross-section with reference to the nitrogen is determined by laboratory measurements in [*Penney and Lapp 1976*]. For incident wavelength of 488nm the cross-section is determined as 0.78×10^{-30} cm^2/sterad, while for nitrogen the value is 0.56×10^{-30} cm^2/sterad. Rotational Raman cross-sections of atmospheric gases are studied in [*Penney et al. 1974*].

4.5. Absorption

In the process of absorption the molecule (or atom) absorbs a photon with energy corresponding to the allowed transition and passes to an exited state. The wavelength dependence of the absorption cross-section of the media is referred as absorption spectrum. The molecular absorption cross-section is a product of the line strength Γ and the spectral shape of the transition Φ,

$$\sigma_{abs}(p_i, T, \nu_0, \nu) = \Gamma(T, \nu_0)\Phi(p_i, T, \nu_0, \nu) \tag{59}$$

Here ν_0 is the central wavenumber of the transition, p_i is for the partial pressures of the absorbing molecule for the ambient constituents.

In the various spectral ranges the absorption of the atmospheric constituents appears as bands and lines, where in combination they determine the atmospheric transparency. This is well seen in the spectrum of the solar radiation reaching the surface (see Figures 1.1 and 1.2 in [*Measures 1984*]).

4.5.1. Absorption in the Near Infrared Range

In the infrared range the line strength may be expressed in the following way [*Bösenberg 2005*]

$$\Gamma(T, \nu_0) = \Gamma(T_0, \nu_0)\left(\frac{T_0}{T}\right)^l exp\left[\frac{E}{k_B}(T_0^{-1} - T^{-1})\right] \tag{60}$$

Here $\Gamma(T_0, \nu_0)$ is the line strength determined at temperature T_0, l is a molecular dependent parameter and E is the energy of the initial state. The spectral shape of the transition is approximated as a Voigt function

$$\Phi(p_i,T,v_0,v) = \frac{\Delta v_C}{\pi \Delta v_D} \int_{-\infty}^{\infty} \frac{exp(-\xi^2)d\xi}{\left(\frac{\Delta v_C}{\Delta v_D}\right)^2 + \left(\frac{v-v_0}{\Delta v_D} - \xi\right)^2} \qquad (61)$$

In Equation (61) Δv_D is the half-width of the spectral line at level $exp(-1)$ of the maximum due to the thermal motion of the molecule. This is named Doppler broadening and has the same nature as the effect described in 4.3. Respectively Δv_C is the half-width of the spectral line at level $exp(-1)$ of the maximum due to the collision with the other molecules, named collision broadening. The parameters Δv_D and Δv_C are expressed as follow

$$\Delta v_D = v_0 \sqrt{\frac{2kT}{c^2 M_x}} \qquad (62)$$

$$\Delta v_C = \Delta v_{C,0}(p_0,T_0)\frac{p}{p_0}\left(\frac{T}{T_0}\right)^\varepsilon \qquad (63)$$

In equations (62, 63), M_x is the mass of the absorbing molecule, $\Delta v_{C,0}(p_0,T_0)$ is the collision half-width at of the spectral line at level $exp(-1)$ for pressure p_0 and temperature T_0, ε and is a parameter.

The shift of the center of the absorption line is another pressure-dependent effect. Considering absorbing constituent in the atmosphere, the dominating effect is the line-shift induced by the air-pressure, expressed in the following way, where η is also a parameter

$$v_0(p,T) = v_0(p_0,T_0) + \frac{p}{p_0}\left(\frac{T}{T_0}\right)^\eta dp \qquad (64)$$

There are important implications from equations (61-64) for the propagation of the laser beam in the atmosphere, containing the absorbing constituent: The line strength, shape and center wavelength for the absorbing constituent depends on the atmospheric temperature and pressure, i.e., on the altitude. Also, the pressure broadening depends on both the partial pressure of the absorbing constituent and the air pressure.

The near infrared spectral range spans from 700nm to 2.5 μm. It contains a multitude of vibrational overtones and combinations of the fundamental vibrational-rotational band of important atmospheric constituents as H_2O, CO_2, CH_4, etc. and other pollution gases. A large number of these lines demonstrate strong absorption and narrow linewidth. (See in [Rothman et al. 2005], in HITRAN database http://www.hitran.com and in European Space Agency, http://badc.nerc.ac.uk/data/esa-wv). This spectral range is very prospective for lidar detection of H_2O, CO_2 and other greenhouse gases, and various pollutions. Also, a strong oxygen absorption band exists around 720nm. Figure 13 presents parts of the absorption spectrum of the water vapour in the near infrared.

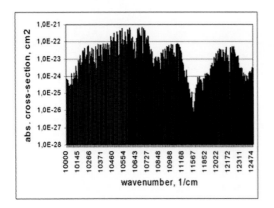 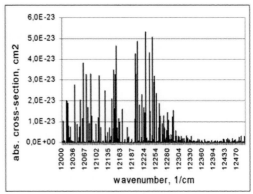

Figure 13. Absorption bands and lines of water vapour in the near IR spectral range. Left: in the range 801.7nm – 1000nm; right: Zoom for the range 801.7nm – 833.3nm. (in accordance to http://www.hitran.com and http://badc.nerc.ac.uk/data/esa-wv).

4.5.2. Absorption in the Ultraviolet and Visible Spectral Range

The absorption in the ultra-violet and in the visible ranges is due to electronic transitions with vibrational-rotational structure, forming large absorption bands. A number of important small constituents show broad absorption bands in this range with complicated and partially overlapping structure: ozone [*Daumont et al. 1992, Malicet et al. 1995*]), SO$_2$ [*Manatt and Lane 1993*], nitrogen oxides [*Schneider et al. 1987*] and other industrial pollution gases [*Schoulepnikoff et al. 1998*].

The absorption spectra of the ozone, sulphur dioxide and nitrogen dioxide are presented in Figure 14. The ozone absorption bands between 200nm and 350nm are known as Hartley and Hugging bands. They form a single broadband continuum. The sulphur dioxide is another molecule showing strong absorption in the UV demonstrating a number of small peaks in the range 285nm-310nm. The nitrogen dioxide absorption band is also broad showing small peaks around 400nm-440nm.

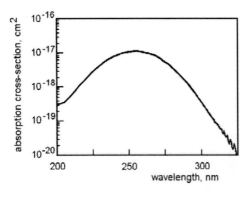 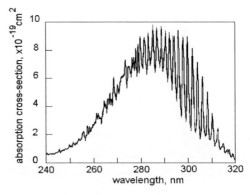

Figure 14 (Continued).

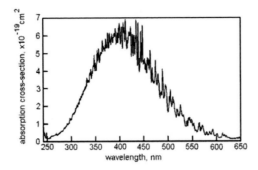

Figure 14. Absorption cross-section of selected minor constituents in the UV-visible spectral range. Top left: Ozone (in accordance to [*Daumont et al. 1992, Malicet et al .1995*]); Top right: Sulfur dioxide (in accordance to [*Manatt and Lane 1993*]). Bottom: Nitrogen dioxide (in accordance to [*Schneider et al. 1987*]).

5. ELASTIC BACKSCATTER LIDAR

The laser wavelength in the elastic backscatter lidar shall not coincide with absorption lines or bands of atmospheric constituents, so we may neglect the absorption in the transmission term. Presently the most used laser in such lidar is the Nd:YAG with its 2nd and 3rd harmonics. There are available detectors for the laser wavelengths (the fundamental and its harmonics) with high quantum efficiency and low noise. The transmitted power, the telescope area, the spectral selection components, the signal acquisition electronics, etc. may be quite different for the different lidar set-ups, depending on the required sensitivity, range of detection and the integration time for a single measurement.

Equation (12) for the elastic backscatter may be written in a simplified form.

$$S(r) = E_\lambda(r)r^2 = C \cdot \beta(r) \cdot e^{-2\int_0^r \alpha(r')dr'} = C.\left[\beta^{aer}(r) + \beta^{mol}(r)\right] \cdot e^{-2\int_0^r \alpha(r')dr'} \qquad (65)$$

Here C is a constant including all factors not depending on the range in Equation (12). The equation is restricted above the full overlap range, i.e., $O(r) \equiv 1$. Equation (65) contains two unknowns: $\beta(r) \cdot$ and $\alpha(r)$. In this way it is not determined and in the general case cannot be solved. Anyway, there are important cases in which we may derive solutions for atmospheric parameters.

5.1. Layers with Low Aerosol Content in Molecular Atmosphere

Let us consider aerosol layers with low aerosol content, where above and/or below the atmosphere is aerosol free. We assume also that the altitudes of interest are above the attenuating layers in the atmosphere (troposphere), i.e., we neglect the attenuation in and between the probed layers. Such conditions may take place in probing stratospheric aerosol layers. For the aerosol free altitudes we may write Equation (65) as

$$S(r_0) = C \cdot \beta^{mol}(r_0)T_{tr}^2 = CN(r_0)\left[\frac{d\sigma_R}{d\Omega}\right]_{\theta=\pi} TR^2 \qquad (66)$$

Here TR is the transmission factor to the probed layers. For the parts of the probed profile with aerosol Equation (66) may be written as

$$S(r) = C \cdot (\beta^{mol}(r) + \beta^{aer}(r))TR^2 \qquad (67)$$

Solving (66) for C, replacing in (67) we obtain

$$\frac{S(r)\beta^{mol}(r_0)}{S(r_0)\beta^{mol}(r)} = \frac{S(r)N(r_0)}{S(r_0)N(r)} = \frac{\beta^{mol}(r) + \beta^{aer}(r)}{\beta^{mol}(r)} = scr(r) \qquad (68)$$

I.e., in the assumed conditions the elastic backscatter lidar equation provides $\beta^{aer}(r)$ or $scr(r)$. The value of $\beta^{mol}(r_0)$ may be obtained from atmospheric models or from nearest meteorological radiosounding.

5.2. Molecular Atmosphere: Determination of the Atmospheric Temperature

Above the layers of background stratospheric aerosol the atmosphere is aerosol-free, i.e., $\beta^{aer}(r) = 0$. The transmission of the atmosphere is determined predominantly by the layers below the altitudes of interest. Then Equation (65) becomes (only for this case, here we replace r with h, for consistency with the references)

$$S(h) = C \cdot \beta^{mol}(h)TR^2 = CN(h)\left[\frac{d\sigma_R}{d\Omega}\right]_{\theta=\pi} TR^2 = C'n_{molec}(h) \qquad (69)$$

Here we combine TR^2 and C in C'. We recall equations (34, 36) and that the average molecular mass of the air is constant (what is valid till the mesopause). We assume signal detection in discrete altitude bins Δh. In each altitude bin we approximate the gravitational acceleration in a linear form as $\tilde{g}(h_i) = [g(h_i) + g(h_i + \Delta h)]/2$. For the average $\bar{n}_{molec}(h_i)$ in each altitude bin, starting from h_i, we take exponential approximation, where $\bar{h}$ is a parameter, i.e.

$$n_{molec}(h_i + \Delta h) = n_{molec}(h_i)exp(-h/\bar{h}), \qquad (70)$$

or

$$\bar{n}_{molec}(h_i) = [n_{molec}(h_i) - n_{molec}(h_i + \Delta h)] / [ln\, n_{molec}(h_i) - ln\, n_{molec}(h_i + \Delta h)] \quad (71)$$

From equations (70 and 71) we may obtain the following relation for the temperature

$$T(h_i) = T(h_i + \Delta h) \frac{n_{molec}(h_i + \Delta h)}{n_{molec}(h_i)} +$$

$$\frac{\tilde{g}(h_i) M_{air}}{k_B n_{molec}(h_i)} [n_{molec}(h_i) - n_{molec}(h_i + \Delta h)] / [ln\, n_{molec}(h_i) - ln\, n_{molec}(h_i + \Delta h)] \qquad (72)$$

From Equation (72) we obtain $T(h_i)$ at successive steps, replacing the values for $n_{molec}(h_i)$ and $n_{molec}(h_i + \Delta h)$ with the detected signal from Equation (69), and starting with reference temperature and density at a reference height (taken typically from atmospheric models). More details of this method may be found in [*Chanin and Hauchecorne 1984, Behrendt 2005*].

5.3. Optically Dense Aerosol Atmosphere

Let us consider probing of layers with high aerosol content and respectively high optical depth. In such case the transmission from the lidar to the probed altitude and between probed layers varies with r.

Let us consider probing of layers with high aerosol content and respectively high optical depth. In such case the transmission from the lidar to the probed altitude and between probed layers varies with r.

We take the natural logarithm of Equation (65) and then the derivative

$$\frac{d\tilde{S}(r)}{dr} = \frac{1}{\beta(r)} \frac{d\beta(r)}{dr} - 2\alpha(r) \qquad (73)$$

Here $\tilde{S}(r) = log(S(r))$. The development requires a specific relation between α and β. As we may find good approximation for β^{mol} and α^{mol}, further we discuss β^{aer} and α^{aer}. To solve the equation for $\beta^{aer}, \alpha^{aer}$ it is necessary to assume a priori knowledge or "best guess" for lr^{aer}. There are two options in solving this problem [*Klett, 1981, Fernald, 1984*].

5.3.1. The Klett's Solution
In this approach the accepted relation between β and α is [*Klett, 1981*].

$$\beta(r) = C'(\alpha(r))^k \qquad (74)$$

Here C' and k are constants (see also [*Fenn, 1966*]). This relation applies only where aerosol backscattering dominates the molecular one. Under such assumption, Equation (73) becomes a Bernoulli equation for α with a mathematically stable solution

$$\alpha(r) = \frac{exp\left(k^{-1}(\tilde{S}(r) - \tilde{S}_f)\right)}{\alpha_f^{-1} + 2\int\limits_{r_f}^{r} dr' exp\left(k^{-1}(\tilde{S}(r') - \tilde{S}_f)\right)} \tag{75}$$

Here r_f is a reference range, $\alpha_f = \alpha(r_f)$ and $\tilde{S}_f = \tilde{S}(r_f)$; r_f is selected to be at the top of the ranges where we are looking for solution. A solution with C' variable with range is suggested also in [*Klett, 1985*].

5.3.2 The Fernald's Solution

Another approach is suggested in [*Fernald et al., 1972, Fernald, 1984*]. With constant lr^{aer} and as $lr^{mol} = \dfrac{8\pi}{3}$ is known (see Chapter 4.1), Equation (73) has the following solution

$$\beta(r) = \frac{S(r)\exp\left(-2(lr^{aer} - lr^{mol})\int\limits_{r_f}^{r} dr'\beta^{mol}(r')\right)}{\dfrac{S(r_f)}{\beta(r_f)} - 2lr^{aer}\int\limits_{r_f}^{r} dr'S(r')\exp\left(-2(lr^{aer} - lr^{mol})\int\limits_{r_f}^{r} dr''\beta^{mol}(r'')\right)} \tag{76}$$

Here r_f is again a reference range with a reference backscattering $\beta(r_f)$. As for the Klett's case, it is recommended to take it in the far range. Numerically, this gives a recurrence scheme for the evaluation of β^{aer}. It applies as well in regions of lower aerosol content where the molecular contribution is dominant.

5.3.3 About the Reference Values

One frequent application of the inversion algorithm is in determination of the aerosol backscatter coefficient in upward direction. Above certain altitudes we may expect that the reference values for the inversion schemes are $\alpha(r_f) = \alpha^{mol}(r_f)$ or $\beta(r_f) = \beta^{mol}(r_f)$, where α^{mol} and β^{mol} may be determined from atmospheric models or meteorological radiosounding. Anyway, aerosols can be present at any tropospheric level. So this approach shall be applied with care to select r_f or to assume some low (background) $\beta^{aer}(r_f)$.

In many cases when the probing is in horizontal or slant directions, we may assume homogeneity of the total aerosol content over some range, i.e., $\beta(r) \approx const$. With this assumption, the gradient term in the right side of Equation (73) is small, and the mean value of the total extinction over such range is then

$$\alpha_{ref} = -\frac{1}{2}\left\langle \frac{d\tilde{S}(r)}{dr} \right\rangle \qquad\qquad\qquad (77)$$

Here the averaging is over the ranges of interest. This method is referred to as slope method. Examples of its application may be found in [*Kunz and Leeuw 1993, Rocadenbosch et al., 2000*].

5.3.4. Numerical Simulations of the Lidar Signal Inversion

The mathematical stability of the solutions of Equation (73) in 5.3.1 and 5.3.2 shall not be misinterpreted as precision of the obtained values for α^{aer} or β^{aer}. That is, a validation of the results is necessary. A relatively straightforward way for such validation is the numerical simulation. The numerical simulations may show the influence of the following factors on the results of the inversion procedure: the algorithm, the selection of reference values (range or altitude, molecular and aerosol reference backscatter coefficients) and the noise components to the detected signal.

The typical steps in the numerical simulations are as follows: the construction of a model aerosol and molecular backscatter coefficients and aerosol lidar ratio; the construction of the lidar signal and noise components; processing of the constructed lidar signal with "guess" reference values; comparing the solution for the aerosol extinction or backscatter with the model values. Some of these steps are illustrated in Figure 15.

5.3.5. Intercomparison of Aerosol Lidar Measurements in EARLINET

Extensive exercises for verification of the lidar systems and the inversion algorithms for a large number of lidar laboratories in Europe were performed in the EU project EARLINET (European Aerosol Research Lidar NETwork) [*Bösenberg and Matthias, 2003*]. The elastic backscatter lidar inversion method is one of the basic methods accepted in this network. As the uncertainties in the application of the respective procedures are well known, as well as the difficulties in its verification, one priority task in the project was the systematic comparison and standardization of the used measurement and processing procedures in the various participating groups.

This is fulfilled by processing numerically constructed signals [*Böckmann et al., 2004*], as well as with intercomparison campaigns of field measurements involving all partners [*Matthias et al., 2004*]. It shall be noted that although the elastic backscatter lidar inversion methods are practiced for long time (as may be seen from the presented references), this was its first systematic large-scale verification. It confirmed the possibility to use the data from various groups in the same database.

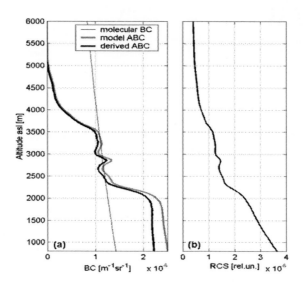

Figure 15. An example of numerical simulation demonstrating the applicability of the lidar signal inversion algorithm in aerosol backscatter determination [*Frioud 2003*]. (a): Model Aerosol and Molecular Backscatter Coefficients, Aerosol Backscatter Coefficient derived by inversion procedure; (b): Range Corrected Signal (logarithm of the product of signal and square of range) calculated from the model Molecular and Aerosol backscatter coefficients and model lidar ratio.

5.4. Depolarisation Ratio

It has been noticed that when the transmitted laser beam is linearly polarised, the depolarisation of the backscattered light from the aerosol and cloud particles is an indication of the particle shape, size with respect to the laser wavelength, refractive index, and aggregation. The most widespread scheme of depolarisation backscatter lidar is with linearly polarized transmitted beam [*Sassen 2005*].

In the depolarisation-backscatter lidar using linearly polarised laser, the receiver is assembled with polarisation analyser, separating the backscatter components having polarisation parallel (p-polarisation) and perpendicular (s-polarisation) to the one of the transmitted beam. Equation (65) may be written separately for the signal at each of the polarisations, where the backscatter coefficients are respectively β_p and β_s. The total volume depolarisation ratio DR^{vol} may be expressed in the following way,

$$DR^{vol}(r) = \frac{\beta_s(r)}{\beta_s(r) + \beta_p(r)} \tag{78}$$

A simplified way to evaluate the depolarisation ratio in cases when the attenuation is the same for the two polarisations or may be neglected is directly from the range corrected backscatter signals.

$$DR^{vol}(r) = \frac{S_{c,s}(r)}{S_{c,s}(r) + \overline{R} \cdot S_{c,p}(r)} \tag{79}$$

Here $\overline{R}$ is the ratio of the total efficiencies of the chain "receiver-detector" for s and p polarizations, i.e., $\overline{R} = K_s \eta_s / K_p \eta_p$, and shall be evaluated independently. In the aerosol science it is used the aerosol depolarisation ratio, defined as

$$DR^{aer}(r) = \frac{\beta_s^{aer}(r)}{\beta_s^{aer}(r) + \beta_p^{aer}(r)} \tag{80}$$

The connections between the various definitions for the depolarisation ratio are summarized in [*Gobbi 1998, Cairo et al. 1999*].

5.5. Determination of Mixing Layer Height

The mixing layer height (MLH, see in Chapter 3) is important for the environmental monitoring and the atmospheric physics. In determining this parameter by backscatter lidar, the aerosol is a tracer, showing the altitudes with homogeneous mixing and with sharp drop of the aerosol concentration. I.e., the exact value of $\beta^{aer}(r)$ is not needed. Compared with the other instruments monitoring of this parameter, the elastic backscatter lidar has the advantages of continuous measurements with high resolution.

The MLH may be determined from the altitude where the lidar signal drops below certain threshold [*Melfi et al. 1985, Steyn et al. 1999*]. Alternative is from the altitude of the minimum of the lidar signal altitude derivative or log-derivative (or "gradient method") [*Menut et al., 1999, Bösenberg and Linné, 2002, Frioud et al. 2003*].

In [*Hooper and Eloranta 1986, Menut et al., 1999*], another method is proposed. As the exchange between the mixing layer and the free troposphere takes place at altitudes of MLH, then this height coincides with the altitude of maximum of the temporal variance of the lidar backscatter signal. Figure 16 presents an example for MLH determination using the lidar signal gradient and variance (from the database used in [*Martucci et al. 2007*]). For comparison, the radiosonde measured temperature profile is presented as well.

5.6. Examples for Elastic Backscatter Lidar Probing in the Stratosphere

The backscatter lidars for stratospheric (and mesospheric) studies employ typically powerful Nd:YAG lasers with second and third harmonics. The typical telescope aperture is between 60cm and 120 cm. Harmonic beamsplitter-filter polychromators are used in the optical receiver to select the backscatter signal at the respective wavelengths. Obligatory is the polarization selection system, since the classification of the clouds requires their depolarisation ratio. Another obligatory block in the lidar set-up is a chopper with the blade positioned at the telescope output and rotation synchronized with the laser pulse. The chopper function is to block the strong backscatter from tropospheric altitudes. Such blocking is necessary to adjust the detector dynamic range to the dynamics of the signal from the altitudes of interest, as well as to avoid afterpulsing. Typical backscatter lidars for stratospheric studies are described in [*Stefanutti et al. 1992, Matthey et al. 1997*].

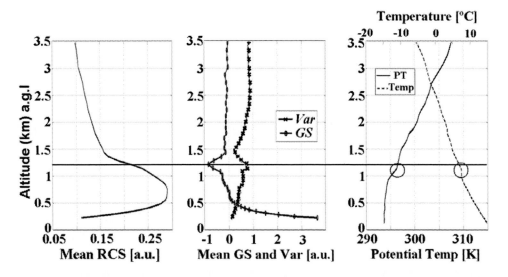

Figure 16. Example illustrating the coincidence of following altitudes: (•) left: altitude of the drop of range-corrected signal value; (•) centre: altitude of minimum of the range-corrected signal gradient (GS) and altitude of maximum of lidar signal variance (Var); (•) right: altitude of planetary boundary layer top from radiosonde temperature profile (circles); (20 May 2002, noon. From the database used in [*Martucci et al. 2007*]).

After the early demonstration of the lidar potential [*Fiocco and Grams 1964*], the backscatter lidar contributed a lot to the stratospheric observation. When the stratospheric aerosol is at its background level, well applicable is the method described in 5.1 for determination of its backscatter coefficient. Anyway, after severe volcanic eruptions the backscatter of the stratospheric aerosol is substantially higher, thus it is necessary to take into account also the extinction of the probed layer. This is helped by the fact that the lr^{aer} of probed stratospheric layers may be either evaluated or "guessed", since they contain predominantly one type of particles.

The stratospheric aerosol impacts the radiative balance in the stratosphere and the chemical processes involved in the reduction of the ozone layer [Hofmann et al., 1994]. Due to this impact the interest in lidar monitoring of the stratospheric aerosol layer is both in its status after volcanic eruptions [*Iwasaka 1986, Thomas et al. 1987, Vaughan et al. 1994, Gobbi 1995*] and in its long-term climatology [*Jäger 2004, Nevzorov et al. 2004*].

The surfaces of the particles of the PSCs are the host for heterogeneous photochemistry process releasing the active chlorine from the CFCs.The way they initiate the chain of chemical reactions leading to the ozone destruction are another object of interest in stratospheric probing. This determines the interest in their observation, where the backscatter lidar play a major role. It shall be noted that the lidars contributed substantially to the classification of the PSC by combined backscatter and polarisation measurements [*Browell et al. 1990, Toon et al. 1990*]. The conditions for PSC appearance are studied in [*Rizi et al.*

1999, Stein et al. 1999, Beyerle et al. 2001, Müller et al. 2004a] and their long-term statistic in [*Blum et al. 2005, di Sarra et al. 2002, Campbell et al. 2006*].

The noctilucent clouds are the highest clouds in the atmosphere [*Thomas 1991*]. They appear at high-latitude and at altitudes of 80-85km, and are visible only in twilight. They consist likely of water ice particles. The backscatter lidar helps to validate the altitudes and latitudes of their appearance [*Thomas et al. 1994, von Cossart et al. 1996*], as well as characteristics of the cloud particles [*von Cossart et al. 1997, von Cossart et al. 1999*].

As shown in Chapter 5.2, the stratospheric temperature at aerosol-free altitudes may be determined from the backscatter lidar signal. The use of Rayleigh backscatter lidar for determination of the temperature at altitudes 30-70km started during 1980s with the motivation to complement the rocket-based sounding [*Hauchecorne and Chanin 1980, Chanin and Hauchecorne 1984*]. Further development of this method made it a primary one in temperature measurements at these altitudes. Continuous lidar measurements made possible the study of atmospheric waves [*Gardner et al. 1989, Marsh et al. 1991, Whiteway and Carswell 1994*] and temperature profile structure [*Sivakumar et al. 2006*]. The accumulation of a long-term record provides the base to evaluate possible climate changes at stratospheric and mesospheric altitudes [*Keckhut et al. 2001, Keckhut et al. 2006*].

5.7. Examples for Elastic Backscatter Lidar Probing in the Troposphere

In studies of transport of tropospheric air, the presence of aerosol layers or its different optical properties are used to indicate air masses with different origin. I.e., in such studies the aerosol backscatter and depolarisation ratios are tracers for the different layers and air mass.

One quite frequent case of long-range aerosol transport is the transport of desert dust. It takes place at tropospheric altitudes and results in an increase of the aerosol load above inhabited areas. The passive optical instruments on satellites may indicate the dust layer but not the altitude profile of its backscatter and depolarisation ratios. Thus, the backscatter lidar, with its capabilities for continuous and altitude resolved aerosol probing, is a key instrument in the study of the long-range transport. Polarisation-backscatter lidar discriminates between dry and wet aerosol in studies of mineral dust [*Gobbi et al. 2000, Gobbi et al. 2003, Ansmann et al. 2003, Immler and Schrems 2003*]. A polarization-backscatter lidar shows also the cloud nucleation effect of the dust particles [*Sassen et al. 2003*]. Crystallized sea-salt and dust particles in the PBL are also studied by depolarisation lidar [*Murayama et al. 1999*].

The organization of lidar network makes it possible to cover with lidar vertical profiling in a coordinated way several sites distributed over a continental scale. Presently such an established lidar network is EARLINET [*Bösenberg et al. 2002*], and more are in the process of establishment [*Murayama et al. 2001, Chaikovsky et al. 2006*]. The advantage of the lidar network EARLINET in investigating Sahara dust transport over Europe was already noted [*Ansmann et al. 2003*].

The backscatter lidar also reveal cases of intercontinental transport of air mass containing aerosol pollutions. Observations taking place over several years are described in [*Kreipl et al. 2000 and Carnuth et al. 2002, Heintzenberg et al. 2003*]. A case study of long-range aerosol transport, observed during a specific wind situation above Central Europe, is presented in [*Mitev et al. 2005*], where Figure 16 presents one observation. It shall be noted that the study of long-range transport of air-mass is possible due to the combination of lidar altitude resolved backscatter profiles and back-trajectory analysis [*Stohl and Seibert 1998*].

Cirrus clouds contain ice-particles and occur in the upper troposphere. Such ice-clouds have different transmission and reflection in the visible and infrared spectral ranges and play an important role in Earth radiation balance. Ultra-thin layers of ice particles appear in the tropics around the tropical tropopause, i.e., up to around 17km. The polarisation measurements make it possible to discriminate between liquid and ice aerosol and in this way to characterise such clouds [Sassen and Liou 1979, Platt et al. 1987, Sassen 2005]. The lidar contributed substantially to the investigation of the cirrus clouds structure, their appearance in various geographic regions [Thomas et al. 1990, Del Guasta et al. 1998], as well as to their statistics [Chen et al. 2002]. Examples of recent studies, where the polarisation lidar participates in synergy with radar other instruments, are presented in [Comstock et al. 2002, Iwasaki et al. 2004, Iwasaki et al. 2006]. Observations are also presented in [Immler and Schrems 2006].

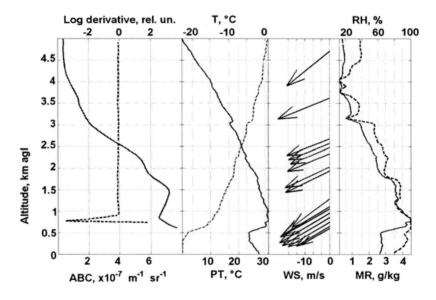

Figure 17. Lidar measurements showing aerosol layer in the free troposphere during Bise event. The 1st panel from left: aerosol backscatter (solid) and log-derivative of range-corrected signal (dashed); The 2nd panel from left: Temperature (T, solid line) and potential temperature (PT, dashed line). The 3rd panel from left: Wind speed (WS). The 4th panel from left: Water vapour mixing ratio (MR, solid line) and relative humidity (RH, dashed line); measurement December 11, 2001, noon.

Another elastic backscatter lidar study in the troposphere, where the aerosol is a tracer, is the vertical transport of aerosol-rich air mass, typically occurring over complex terrain. Such process leads to an enhanced PBL-troposphere exchange and in this way to potential for long-range pollution transport [*Carnuth and Trickl 2000, Frioud et al. 2003*].

The elastic backscatter lidar measures in a continuous way, which makes it very useful in observations of temporal development in the planetary boundary layer. In such measurements the objective is usually MLH, its daily cycle or its variation due to the synoptic conditions. Widespread applications find useful the methods based on the gradient of the range-corrected lidar signal [Frioud et al. 2004, Sicard et al. 2006, Wiegner et al. 2006]. Figure 18 presents one example for diurnal variation of MLH observed by elastic lidar during the project BUBBLE above Basel [Rotach et al. 2005].

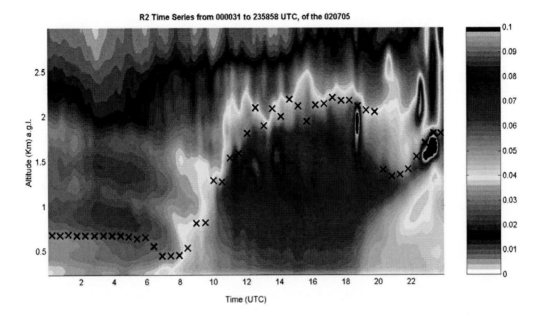

Figure 18. Elastic lidar measurement during one PBL daily cycle (2nd July 2002) above the centre of Basel; the grey-scale presents the lidar range corrected signal; the MLH determined by the gradient method is marked with "x".

5.8. Backscatter Lidar Measurements with High-Spectral Resolution

Due to the Doppler broadening the spectral profile of the Rayleigh (molecular) backscatter is distributed over the range of 0.2cm^{-1} (Figure 10). As the mass of the aerosol particle is several orders of magnitude larger than the mass of the molecule of the atmospheric gases, the Doppler broadening of the aerosol backscatter is a lower order of magnitude less than the same for the molecular backscatter (what may be seen from Equation (49), replaing the molecular mass with the aerosol one). This difference in the broadening of the molecular and the aerosol backscatter defines the principle of the high-spectral resolution lidar: to select spectrally the aerosol and the molecular components in the elastic backscatter and to detect them separately. In this way Equation (65) is divided to two equations, one for only the molecular component and one for predominantly the aerosol component. For simplicity below we neglect the small molecular contribution in such aerosol channel.

$$S^{mol}(r) = C^{mol}.\beta^{mol}(r).e^{-2\int_0^r \alpha(r')dr'} \tag{81}$$

$$S^{aer}(r) = C^{aer}.\beta^{aer}(r).e^{-2\int\limits_{0}^{r}\alpha(r')dr'} \tag{82}$$

The ratio of Equation (81) and Equation (82) yields the following

$$\left[S^{aer}(r) \Big/ S^{mol}(r) \right] = \left[C^{aer}.\beta^{aer}(r) \Big/ C^{mol}.\beta^{mol}(r) \right] \tag{83}$$

After a calibration for the ratio C^{aer}/C^{mol} and independent evaluation of $\beta^{mol}(r)$ from the molecular density, we may determine $\beta^{aer}(r)$.

The technical problems with the realisation of the high-spectral resolution lidar are inherent in using powerful laser having ultra-narrow-line and the ultra-narrow-line spectral filter coupled to the laser line [*Shipley et al. 1983*]. Fabry-Perot interferometers are typically used, as well as atomic and molecular absorption filters. A review of the high-spectral resolution lidars and applications, as well as the spectral selection techniques, are presented in [*Eloranta 2005*].

Another application of the high-spectral resolution lidar technique is the wind velocity measurement with the direct detection Doppler wind lidar. In this technique, two parts in the Doppler broadened molecular backscatter line are selected by ultra-narrow filters and are separately detected. These two parts are symmetrical with respect to the line center, i.e., one part is with a positive wavelength shift while the other is with a negative one. When the backscattering part of the atmosphere moves with respect to the lidar, this causes a Doppler (wind) shift of the Doppler-broadened backscatter line. The maximum of the line moves further from one of the filters closer to the other. In this way the ratio of the signals detected through the filters depends on the Doppler shift, i.e., on the wind velocity [*Werner 2005*].

6. RAMAN BACKSCATTER LIDAR

6.1. The Raman Lidar Equations

Let us write Equation (11) for the Raman backscatter signal from constituent M in the following form

$$E_R(r) = E_L K \eta_R O(r) \frac{A}{r^2} \frac{c\tau}{2} \beta_R(r) exp\left(-\int\limits_{0}^{r}[\alpha_\lambda(r') + \alpha_R(r')]dr' \right) \tag{84}$$

The Raman volume backscatter coefficient is determined as

$$\beta_R(r) = N_M(r) \left[\frac{d\sigma_{Ram}(v)}{d\Omega} \right]_{\theta=\pi} \tag{85}$$

Here $\left[\dfrac{d\sigma_{Ram}(v)}{d\Omega}\right]_{\theta=\pi}$ stays either for $\left[\dfrac{d\sigma_{vib}(v)}{d\Omega}\right]_{\theta=\pi}$ or for $\left[\dfrac{d\sigma_{rot}(J)}{d\Omega}\right]_{\theta=\pi}$;

$N_M(r)$ is the molecular number density of the respective constituent (this may be nitrogen, oxygen, water, etc.). Let us also note that the round-trip transmission is determined by two extinction coefficients: one at laser wavelength (subscript λ) and one is at the Raman wavelength (subscript R).

Due to the low value of the Raman cross-sections and thus of the backscatter signal, the Raman lidars require lasers with high average power, receivers with larger aperture and efficient detectors. The laser is a key part of the Raman lidar. The first Raman lidar demonstrations used low-power Ruby-lasers with wavelength 694nm (fundamental) or 347nm (second harmonic), or Cu-vapour laser (510.6nm). The low average power limited the measurement ranges to 1-2 km. The availability of lasers with high pulse energy and high pulse repetition rate made possible applications covering the troposphere and lower stratosphere. Presently the typical Raman backscatter lidar uses the second and third harmonics of the Nd:YAG laser (respectively at 532nm and 355nm), as well as excimer lasers XeCl at 308nm and XeF at 351nm.

The requirements for the spectral selection in the Raman lidar are much stronger than in the elastic backscatter lidar. The spectral transmission shall be narrow-band, from 1-2nm to parts of nm and with high transmission. When interference filters are used, the transmission may be as high as 50%-60%. Particularly strong is the requirement for out-of-line suppression in order to reduce the intrusion of the elastic backscatter: this is not less than 8 to 10 orders of magnitude. In the present advanced Raman lidars the spectral selection of the Raman backscatter is performed by interference filters combined with polychromators. The polychromators are based on a combination of dichroic beamsplitters, spectrometers and etalons. The detection is typically performed by photomultipliers in photon counting mode.

6.2. Determination of Extinction and Backscatter Coefficients

Let us start with Equation (84) for the vibrational Raman backscatter of nitrogen, assuming $O(r) = 1$ and combining the instrument parameters not depending on the range in one constatnt K_{N2}

$$S_{N2}(r)=E_{N2}(r)r^2 = K_{N2}N_{N2}(r)\left[\frac{d\sigma_{N2}(v)}{d\Omega}\right]_{\theta=\pi} exp\left(-\int_0^r [\alpha_{las}(r')+\alpha_{N2}(r')]dr'\right) \quad (86)$$

The subscript "las" and "$N2$" are for the value of the laser and nitrogen Raman wavelength, respectively. Further, we transform Equation (86) in the following way: we take the natural logarithm and then the derivative with respect to the range. Considering no altitude dependence of the Raman scattering cross–section, we obtain the following relation

$$\frac{d}{dr}\left\{ln\left[N_{N2}(r)/S_{N2}(r)\right]\right\} = \alpha_{las}(r)+\alpha_{N2}(r) \quad (87)$$

The Raman backscatter signal provides the mean value of the extinction coefficient in the wavelength range between the laser and the Raman shifted wavelength. In case these two wavelengths are very close to each other, as it is the case of pure rotational Raman scattering, then this corresponds to the extinction coefficient at the laser wavelength.

Accounting for equations (14, 15) we derive the following:

$$\frac{d}{dr}\left\{ln\left[N_{N2}(r)/S_{N2}(r)\right]\right\} - \alpha_{las}^{mol}(r) - \alpha_{Ram}^{mol}(r) = \alpha_{las}^{aer}(r) + \alpha_{Ram}^{aer}(r) \quad (88)$$

or

$$\frac{d}{dr}\left\{ln\left[N_{N2}(r)/S_{N2}(r)\right]\right\} - \alpha_{las}^{mol}(r)\left[1 + \left(\frac{\lambda_{las}}{\lambda_{N2}}\right)^4\right] = \alpha_{las}^{aer}(r)\left[1 + \left(\frac{\lambda_{las}}{\lambda_{N2}}\right)^{a'(r)}\right] \quad (89)$$

In Equation (89) $a'(r)$ is the Ångström exponent in the wavelength dependence of the $\beta^{aer}(r)$. With known altitude distribution of the molecular density (either from atmospheric models or from meteorological radiosounding) the Raman lidar determines the mean value of the aerosol extinction coefficient between the laser and the Raman wavelengths. In case we may make a good assumption for the Ångström exponent, it determines also the aerosol extinction at any of these wavelengths. For practical applications Equation (89) shall be written in discreet (finite) range resolution.

The elastic backscatter coefficient $\beta_\lambda(r) = \beta_\lambda^{aer}(r) + \beta_\lambda^{mol}(r)$ may be determined from the simultaneously detected Raman and elastic backscatter signals. Unlike the methods described in Chapter 5, this determination is without assumptions for the $lr^{aer}(r)$. Assuming the same overlap function for the elastic and Raman receivers, equations (12, 84) may be written in the following forms, where the subscripts "λ" and R are respectively for the elastic and Raman backscatters. We again combine the instrument parameters not depending on the range in one constatnt in each equation, respectively K_λ and K_R.

$$E_\lambda(r) = K_\lambda(\beta_\lambda^{aer}(r) + \beta_\lambda^{mol}(r))exp\left(-2\int_0^r[\alpha_\lambda(r')]dr'\right) \quad (90)$$

$$E_R(r) = K_R\beta_R(r)exp\left(-\int_0^r[\alpha_\lambda(r') + \alpha_R(r')]dr'\right) \quad (90a)$$

The ratio of equations (90, 90a) yields the following

$$\frac{E_\lambda(r)}{E_R(r)} = \frac{K_\lambda}{K_R}\frac{\beta_\lambda(r)}{\beta_R(r)}exp\left(-\int_0^r[\alpha_\lambda(r') - \alpha_R(r')]dr'\right) \quad (91)$$

Let us assume that at range (or altitude) r_{ref} the atmosphere is aerosol-free, i.e., $\beta_\lambda(r_{ref}) = \beta_\lambda^{mol}(r_{ref})$. Then the Equation (91) written for range r_{ref} gives

$$\frac{E_\lambda(r_{ref})}{E_R(r_{ref})} = \frac{K_\lambda}{K_R} \frac{\beta_\lambda^{mol}(r_{ref})}{\beta_R(r_{ref})} exp\left(-\int_0^{r_{ref}}[\alpha_\lambda(r')-\alpha_R(r')]dr'\right) \qquad (92)$$

Taking the ratio of equations (91, 92) we obtain for $\beta_\lambda(r)$ the following expression

$$\beta_\lambda(r) = \beta_\lambda^{aer}(r) + \beta_\lambda^{mol}(r) = \frac{E_\lambda(r)E_R(r_{ref})}{E_R(r)E_\lambda(r_{ref})} \frac{\beta_\lambda^{mol}(r_{ref})\beta_R(r)}{\beta_R(r_{ref})} exp\left(-\int_r^{r_{ref}}[\alpha_\lambda(r')-\alpha_R(r')]dr'\right) (93)$$

In addition to the detected elastic and Raman backscatter, to perform a step-wise procedure, we need the profile of the atmospheric density from a model or from meteorological radiosonding. After determining the values of $\alpha_\lambda^{aer}(r)$ from Equation (89) and of $\beta_\lambda^{aer}(r)$ from Equation (93), we may also obtain the lidar ratio $lr(r)$. I.e., from the combined elastic-Raman measurements we may deduce if the aerosol type changes with the altitude.

The method to determine the atmospheric extinction and backscatter by the Raman backscatter of the vibrational line of nitrogen were analysed in [*Ansmann et al. 1990, Ansmann et a.l 1992, Whiteman et al. 1992*]. Advanced applications motivated the realisation of robust Raman lidar hardware and studies of the uncertainties of the measurements [*Ferrare et al. 1998, Ferrare et al. 1998a*]. In [*Wandinger 1998*] it is demonstrated how to account for the multiple scattering in Raman extinction and backscatter measurements.

As we have seen, the Raman and the combined Raman-elastic lidars provide independent determination of $\alpha^{aer}(r)$, $\beta^{aer}(r)$, as well as $lr^{aer}(r)$. The use of the elastic and Raman backscatters from more than one probing wavelength allows the determination also of the wavelength dependences of these parameters [*Ferrare et al. 1998, Müller et al. 1999, Ansmann and Müller, 2005*]. As shown there, with the independently determined aerosol extinction and backscatter for more than one wavelength, it may be possible to recover integral properties of the particle ensemble as the surface-area-weighted mean radius, total surface area concentration and total volume concentration.

The Raman lidar is a necessary advanced sensor in aerosol lidar networks. This motivated a comparison of the results from signal processing algorithms in the lidar network EARLINET [*Pappalardo et al. 2004*]. The results from this comparison demonstrate the possibility to have a unified database from the different stations using Ramah lidars for extinction measurements. In the frame of EARLINET, several Raman lidars performed a coordinated study of the vertical aerosol distribution and the modifications of the air masses during their transport over Europe [*Matthias et al. 2004, Wandinger et al. 2004*]. The Raman lidar provides a contribution to the long-term routine measurements in the PBL and troposphere [*Matthias et al. 2004, De Tomasi et al. 2006*] as well as observations of long-range aerosol transport [*Müller et al. 2004*].

The progress in the laser and detection techniques makes possible the participation of Raman lidars in field campaigns. During the Aerosol Characterisation Experiment, the Raman lidar identified layers of non-absorbing continental pollution [*Müller et al. 2002*]. During the campaign INDOEX [*Franke et al. 2001*] it detected episodes of anthropogenic particle transport from Indian sub-continent to the tropical region of the Indian Ocean.

Due to the lower level of the Raman backscatter compared to the elastic one, the Raman lidar is typically applied in tropospheric studies. Nevertheless, there are examples of successful stratospheric measurements of aerosol optical properties [*Ferrare et al. 1992, Wandinger et al. 1995*].

The works quoted in this paragraph deal with measurements of the elastic aerosol extinction and backscatter using vibrational Raman backscatter from nitrogen. The use of the pure rotational Raman backscatter cancels the necessity for wavelength correction in equations (88-89). On the other hand, the spectral selection of the pure rotational Raman spectra is more demanding, where care shall also be taken for the temperature sensitivity of the rotational Raman backscatter power [*Arshinov et al. 1983*]. Determination of atmospheric extinction using pure rotational Raman scattering from nitrogen and oxygen is presented in [*Mitev et al. 1990, Balin et al. 2004*], while a comparison between this method and the Klett inversion procedure in elastic backscatter lidar is presented in [*Mitev et al. 1992*].

6.3. Raman Lidar Measurements of the Concentration of Atmospheric Constituents

Let us consider the Raman backscatter signal from molecular constituent x with number density $N_X(r)$ and from nitrogen:

$$E_X(r) = E_L K_X \eta_X O(r) \frac{A}{r^2} \frac{c\tau}{2} N_X(r) \left[\frac{d\sigma_X(v_X)}{d\Omega} \right]_{\theta=\pi} exp\left(-\int_0^r [\alpha_{las}(r') + \alpha_X(r')] dr' \right) \quad (94)$$

$$E_{N2}(r) = E_L K_{N2} \eta_{N2} O(r) \frac{A}{r^2} \frac{c\tau}{2} N_{N2}(r) \left[\frac{d\sigma_{N2}(v_{N2})}{d\Omega} \right]_{\theta=\pi} exp\left(-\int_0^r [\alpha_{las}(r') + \alpha_{N2}(r')] dr' \right) \quad (94a)$$

Let us assume identical overlap functions for the two Raman receiving channels and range independent Raman scattering cross-sections (i.e., both temperature and pressure independent). Then the ratio of equations (94, 94a) and subsequent transformations give the following expression for $N_X(r)$:

$$N_X(r) = C N_{N2}(r) \left[\frac{E_X(r)}{E_{N2}(r)} \right] exp\left(-\int_0^r [\alpha_{N2}(r') - \alpha_X(r')] dr' \right) \quad (95)$$

$$\text{Where } C = \left(\frac{K_{N2}\eta_{N2}}{K_X\eta_X}\right)\left\{\frac{\left[\dfrac{d\sigma_{N2}(v_{N2})}{d\Omega}\right]_{\theta=\pi}}{\left[\dfrac{d\sigma_X(v_X)}{d\Omega}\right]_{\theta=\pi}}\right\} \tag{96}$$

To obtain the mixing ratio with nitrogen of the constituent x, $N_X(r)$, we need to know the following parameters: the ratio of the Raman scattering cross-sections, the ratio of the efficiencies for the two measurement channels, as well as the difference between the extinction coefficients at the Raman wavelength for x and for nitrogen. Since the difference of the molecular extinctions may be evaluated from the atmospheric density, the unknown is the difference between the aerosol extinction coefficients. In cases when the measurements are in conditions of higher aerosol optical depth, the contribution of this term to the systematic error needs to be taken into account. A source of instrumental systematic error is also the width of the spectral selection of the vibrational Raman band. A selection that cuts temperature-dependent lines in the rotational-vibrational structure will lead to altitude-dependent systematic uncertainty [*Bribes et al. 1976, Whiteman et al. 1993, Whiteman 2003*].

The importance of the water vapour and its relatively high concentration motivated early interest to water vapour Raman lidars [*Melfi et al. 1969, Cooney 1970, Melfi 1972*]. At that early stage Raman lidar measurements were just a demonstration of the feasibility. Successive measurements showing the dynamics of the water vapour altitude profile at tropospheric altitudes are reported in [*Melfi and Whiteman 1985*].

The measurements were extended to high troposphere and tropopause altitude in [*Vaughan et al 1988*]. In this study, the determination of the calibration constant C in Equation (96) includes a precise determination of the sensitivity ratio of the two measurement channels and spectroscopy data for the Raman scattering cross-sections of the nitrogen and the water vapour. The obtained results were compared with radiosondes showing coincidence within the statistical error.

An alternative way to evaluate the constant C in Equation (96) is by a calibration of the lidar measurements with measurements from traditional *in situ* sensors. A description of such calibration in the case of water vapour Raman lidar is presented in [*Sherlock et al. 1999, Sherlock et al. 1999a*].

With the progress of laser, spectral filter and detector technique, a new generation of water vapour Raman lidars have been realised in a number of groups, sometimes combining the water vapour with Raman extinction and elastic backscatter measurements [*Goldsmith et al. 1998, Althausen et al. 2000, Turner et al. 2002, Whiteman 2003a*]. The addition of extinction measurements brings the possibility for a reliable correction of the systematic errors due to the aerosol extinction.

Recently, a new generation of Raman lidars find application in field experiments, together with a number of complementary sensors. During such field campaigns a validation of the Raman lidar measurements with different types of radiosondes has been performed as well [*Ferrare et al. 1995*]. A recent and representative example of the potential of the water vapour Raman lidar in such campaigns may be see in [*Whiteman et al. 2006, 2006a and 2006b*]. A systematic review of the present maturity of the Raman lidar measurements of water vapour is presented in [*Wandinger 2005*].

One perspective in the Raman Lidar technique is the observation of the cloud liquid water [*Melfi et al. 1997, Vesselovskii et al. 2000*]. A limitation in such measurements is the attenuation in the cloud, where the typical penetration depth is in the order of 100-200m. Another problem is that the Raman scattering cross–section of the small droplet is not the same as for the bulk water: it has a resonance character, depending on the wavelength and the droplet diameter. A solution to this problem is proposed in [*Veselovskii et al. 2002*]. A third problem is the separation of the water spectra of the droplets and the vapour. A solution may be the high-resolution spectral selection [*Arshinov et al. 2002*]. At last, there is strong temperature dependence in the liquid water spectrum, which requires a selection of the part of the spectrum showing no or low temperature sensitivity [*Whiteman et al. 1999*]. A proposal to use the Raman scattering of ice in lidar measurement of cirrus clouds is presented in [*Wang et a. 2004*].

There are few examples of Raman lidar measurements of other minor atmospheric constituents. The reason is that such measurements are very demanding, due to the low-level Raman signal combined with sometimes high aerosol optical depth (in the case of industrial pollution). With the advance of Raman lidar spectral selection and detection, such difficulties may be solved. An example showing the high technical level of Raman lidar systems for pollution measurement is presented in [*Arshinov et al. 1996*].

6.4. Rotational Raman Lidar for Atmospheric Temperature Measurement

The pure rotational Raman spectra of nitrogen and oxygen have two parts, where the temperature derivative of the cross-section has different signs (see Chapter 4.4 and Figure 19). We assume that the Raman lidar receives independently two groups of pure rotational Raman lines of both oxygen and nitrogen, where the temperature derivatives of the cross-sections are with opposite signs. The equations for the received signals may be expressed as follows:

$$E_{R1}(r) = E_L K_{R1} \eta_{R1} O(r) \frac{A}{r^2} \frac{c\tau}{2} N \ (r) \left[\frac{d\sigma \ (JR1)}{d\Omega} \right]_{\theta=\pi} TR1(r) \qquad (97)$$

and

$$E_{R2}(r) = E_L K_{R2} \eta_{R2} O(r) \frac{A}{r^2} \frac{c\tau}{2} N \ (r) \left[\frac{d\sigma \ (JR2)}{d\Omega} \right]_{\theta=\pi} TR2(r) \qquad (97a)$$

where $N(r)$ is the air number density. We define the transmission factors as

$$TR1(r) = exp \left(- \int_0^r \left[\alpha_\lambda(r') + \alpha_{R1}(r') \right] dr' \right) \qquad (98)$$

and

$$TR2(r) = exp\left(-\int_0^r [\alpha_\lambda(r') + \alpha_{R2}(r')]dr'\right)$$ (98a)

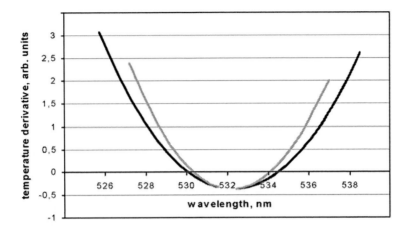

Figure 19. Temperature derivative of the pure rotational Raman cross-sections of nitrogen and oxygen. Black line: nitrogen. Grey line: oxygen.

The combined rotational Raman cross-sections are expressed as the following sums:

$$\sum RR1(r) = \sum_{JR1,N2} t_{JR1}\varepsilon_{N2}\left[\frac{d\sigma_{N2}(JR1)}{d\Omega}\right]_{\theta=\pi} + \sum_{JR1,O2} t_{JR1}\varepsilon_{O2}\left[\frac{d\sigma_{O2}(JR1)}{d\Omega}\right]_{\theta=\pi}$$ (99)

and

$$\sum RR2(r) = \sum_{JR2,N2} t_{JR2}\varepsilon_{N2}\left[\frac{d\sigma_{N2}(JR2)}{d\Omega}\right]_{\theta=\pi} + \sum_{JR2,O2} t_{JR2}\varepsilon_{O2}\left[\frac{d\sigma_{O2}(JR2)}{d\Omega}\right]_{\theta=\pi}$$ (99a)

Here the values ε_{N2} and ε_{O2} are the percentages of the nitrogen and oxygen in air, and the factors t_{JR1} and t_{JR2} are the transmissions of the spectral selection system for the respective rotational Raman lines.

For ranges above the full overlap the ratio of equations (97, 97a) yields

$$\frac{E_{R1}(r)}{E_{R2}(r)} = K\left[\frac{\sum RR1(r)}{\sum RR2(r)}\right] = RT(T)$$ (100)

Here we use the condition $\lambda R1 \approx \lambda R2$ and, following from it, $TR1(r) \approx TR2(r)$. K is the ratio of the efficiencies $K = K_{JR1}\eta_{JR1}(K_{JR2}\eta_{JR2})^{-1}$.

One way to derive the temperature from Equation (100) is by precise spectroscopic measurements of the ratios of the transmissions K_{JR1}, K_{JR2} and the efficiencies η_{JR1}, η_{JR2}.

Another way for temperature derivation is used in [*Arshinov et al. 1983*]. Equation (100) may be approximated as

$$RT(T) = \left[\frac{E_{R1}(r)}{E_{R2}(r)} \right] = K \left[\frac{\sum R1(r)}{\sum R2(r)} \right] = exp(a - b/T) \qquad (101)$$

Here a and b may be determined by calibrating the lidar with *in situ* temperature measurements. In small probed altitudes, i.e., where the interval of temperature changes is also small, there may be even a linear approximation of Equation (101).

The temperature lidar measurements based on rotational Raman scattering was proposed in [*Cooney 1972*]. Due to the technical challenges, for a considerable period the rotational Raman lidar temperature measurements were only demonstrations limited to the PBL and the lower troposphere [*Cooney and Pina 1976, Cohen et al. 1976, Gill et al. 1979, Arshinov et al. 1983, Mitev et al. 1988*]. These early works demonstrated the potential of the method in temperature measurement at altitudes where the aerosol is present and verified the technical requirements of this method.

With more advanced lasers and interference filters the rotational Raman lidar techniques was demonstrated to altitudes of upper troposphere and lower stratosphere [*Vaughan et al. 1993*]. The temperature dependence of the ratio in Equation (100) was determined by precise measurements of the transmission functions of the interference filters selecting the two parts of the rotational Raman spectra, as well as the ratio of the detection channels efficiencies. The obtained results were compared with radiosondes launched from the same site, showing coincidence within the statistical error (Figure 20). Continuous measurements showing the temperature dynamics in the troposphere were also performed.

The rotational Raman temperature measurement is one of the most demanding atmospheric lidar applications. A high-output power and reliable laser is used, typically the second or third harmonics of the Nd:YAG laser, with preferences for the second [*Behrendt 2005*]. The spectral selection system shall provide a sufficient blocking of the high elastic backscatter, just at 1-3nm from the rotational Raman line. A small leakage of the elastic backscatter into the selected rotational bands may compromise the temperature determination. An evaluation of the required blocking is presented also in [*Behrendt 2005*] with the conclusion that such blocking shall be no less than seven orders of magnitude.

The progress in lasers and interference filters makes possible the realisation of advanced rotational Raman lidars. This allows the measurement of rotational Raman lidar temperature at high-altitudes and even in clouds [*Behrendt and Reichardt 2000*]. For additional blocking of the elastic backscatter and the optical background, interferometer-based spectral selection is used, allowing daytime temperature profiling in the troposphere [*Arshinov and Bobrovnikov 1999, Bobrovnikov et al. 2002, Arshinov et al. 2005*]. The implementation of this technique required solving the coupling of the interferometer to the lidar receiver for non-distorted range-resolved measurements [*Serikov et al. 2002, Arshinov et al. 2004*].

Unlike the lidar based on the Rayleigh backscatter (described in Chapter 5) the Rotational Raman lidar is capable of determining the temperature at altitudes where aerosol is present. The combination of these two lidar methods in one instrument allows atmospheric temperature profiling at altitudes from the troposphere to the mesosphere [*Behrendt et al. 2004*]. Such combination also allows temperature measurements at the altitudes of PSC appearance [*Fierli et al. 2001*]. It shall be noted that extension of the measurements to the stratosphere was presented first in [*Nedeljkovic et al. 1993*]. The rotational Raman lidar has been also combined in one instrument with Raman lidar for detection of water vapour concentration and with elastic lidar. In this way it is possible to obtain together the altitude profiles of the temperature and relative humidity profiles [*Matthias et al. 2002, Behrendt et al. 2002, Balin et al. 2004, Balin et al. 2004a*].

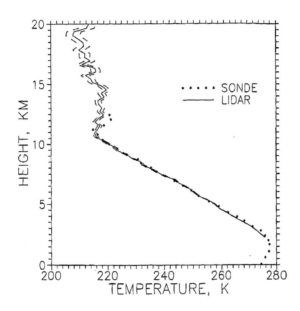

Figure 20. Comparison between a lidar temperature profile measured above Aberythwyth (UK) between 22GMT on 11 December 1991 and 0200 on the 12th and a radiosonde launched also from Aberystwith at 1100 on the 12th. Dashed curves on either side of the lidar profile denote 1σ precision limits; from [*Vaughan et al. 1993*]; Reproduction is with the permission and courtesy of OSA.

7. DIFFERENTIAL-ABSORPTION LIDAR (DIAL)

7. 1. Elastic DIAL

In the elastic DIAL (DIfferential-Absorption Lidar) the lidar probes with two laser beams having different wavelengths. One of these wavelengths coincides with the center of an absorption line (or band) of a selected atmospheric constituent and is referred to as "on"-wavelength, λ_{on}. The other wavelength does not coincide with absorption lines and is referred to as "off"-wavelength, λ_{off}. When the absorption band is broad, there may be still absorption at λ_{off}, but it shall be significantly smaller the absorption at λ_{on}. Let us write the elastic lidar equation for the case of two backscatter signals $E_{on}(r)$ and $E_{off}(r)$ respectively at λ_{on} and λ_{off}. The subscripts "on" and "off" are for the parameters respectively at λ_{on} and λ_{off}.

$$E_{on,}(r) = E_{Lon}K_{on}\eta_{on}O_{on}(r)A\frac{c\tau}{r^2}\beta_{on}(r)exp\left\{-2\int_0^r\{\alpha_{on}(r')+N_X(r)\sigma_{on}^X(r')\}dr'\right\} \quad (102)$$

$$E_{off}(r) = E_{Loff}K_{off}\eta_{off}O_{off}(r)A\frac{c\tau}{r^2}\beta_{off}(r)exp\left\{-2\int_0^r\{\alpha_{off}(r')+N_X(r)\sigma_{off}^X(r')\}dr'\right\} \quad (102a)$$

Here $\sigma_{on}^X(r')$ and $\sigma_{off}^X(r')$ are the molecular absorption coefficients at λ_{on} and λ_{off}; $N_X(r)$ is the number density of the measured constituent.

We make the following successive transformations on equations (102, 102a): multiplying by r^2, taking the natural logarithm and then the r- derivative. As a next step after the transformations, we take the difference of the two equations, resulting in

$$\frac{d}{dr}\left[ln\frac{S_{off}(r)}{S_{on}(r)}\right] = \frac{d}{dr}\left[ln\frac{\beta_{off}(r)}{\beta_{on}(r)}\right] - 2\left[\alpha_{off}(r)+\sigma_{on}^X(r)N_X(r)-\alpha_{on}(r)-\sigma_{on}^X(r)N_X(r)\right] \quad (103)$$

where $E_{on}(r)r^2 = S_{on}(r)$ and $E_{off}(r)r^2 = S_{off}(r)$. From Equation (103) we may obtain $N_X(r)$ as

$$N_X(r) = \left[2\left(\sigma_{on}^X - \sigma_{off}^X\right)\right]^{-1}\frac{d}{dr}\left[ln\frac{S_{off}(r)}{S_{on}(r)}\right] - \left[2\left(\sigma_{on}^X - \sigma_{off}^X\right)\right]^{-1}\frac{d}{dr}\left[ln\frac{\beta_{off}^{mol}(r)+\beta_{off}^{aer}(r)}{\beta_{on}^{mol}(r)+\beta_{on}^{aer}(r)}\right] + \quad (104)$$

$$\left(\sigma_{on}^X - \sigma_{off}^X\right)^{-1}\left[\left(\alpha_{off}^{mol}(r)-\alpha_{on}^{mol}(r)\right)+\left(\alpha_{off}^{aer}(r)-\alpha_{on}^{aer}(r)\right)\right]$$

Transforming Equation (104) in discreet range resolution, we reach the expression

$$N_X(r) = \frac{1}{(\sigma_{on}^X - \sigma_{off}^X)2\Delta r}\left[ln\frac{S_{off}(r+\Delta r)S_{on}(r)}{S_{on}(r+\Delta r)S_{off}(r)}\right] -$$

$$\frac{1}{(\sigma_{on}^X - \sigma_{off}^X)2\Delta r}\left[ln\frac{\{\beta_{off}^{mol}(r+\Delta r)+\beta_{off}^{aer}(r+\Delta r)\}\{\beta_{on}^{mol}(r)+\beta_{on}^{aer}(r)\}}{\{\beta_{on}^{mol}(r+\Delta r)+\beta_{on}^{aer}(r+\Delta)\}\{\beta_{off}^{mol}(r)+\beta_{off}^{aer}(r)\}}\right] + \quad (105)$$

$$\frac{1}{(\sigma_{on}^X - \sigma_{off}^X)}\left[\left(\alpha_{off}^{mol}(r)-\alpha_{on}^{mol}(r)\right)+\left(\alpha_{off}^{aer}(r)-\alpha_{on}^{aer}(r)\right)\right]$$

If $\lambda_{off} \approx \lambda_{on}$, then $\beta_{off}^{mol}(r) \approx \beta_{on}^{mol}(r)$, $\beta_{off}^{aer}(r) \approx \beta_{on}^{aer}(r)$, $\alpha_{off}^{mol}(r) \approx \alpha_{on}^{mol}(r)$ and

$\alpha_{off}^{aer}(r) \approx \alpha_{on}^{aer}(r)$, and Equation (104) allows a straightforward determination of $N_X(r)$,

$$N_X(r) = \frac{1}{\left(\sigma_{on}^X - \sigma_{off}^X\right)2\Delta r}\left[ln\frac{S_{off}(r+\Delta r)S_{on}(r)}{S_{off}(r)S_{on}(r+\Delta r)}\right] \qquad (106)$$

When $\lambda_{off} \neq \lambda_{on}$, to determine $N_X(r)$ from Equation (104) we need to know not only $\sigma_{on}^X - \sigma_{off}^X$, but also a number of correction terms containing aerosol and molecular extinction and backscatter coefficients. The molecular extinction and backscatter coefficients may be determined from the atmospheric density (i.e., from the atmospheric models or from meteorological radiosonding). Nevertheless, the influence of $\alpha_{off}^{aer}(r), \alpha_{on}^{aer}(r)$ and $\beta_{off}^{aer}(r), \beta_{on}^{aer}(r)$ remains. Particularly high is the contribution of the term containing $\beta_{off}^{aer}(r), \beta_{on}^{aer}(r)$ in the PBL due to the high gradient of the aerosol content in this part of the atmosphere.

Figure 21 presents calculated backscatter signal profiles at "on" and "off" wavelengths for water vapour DIAL. In this simulation we assume PBL till 1200m containing aerosol and water vapour, and a layer containing water vapour but no aerosol from 2000 to 3000m.

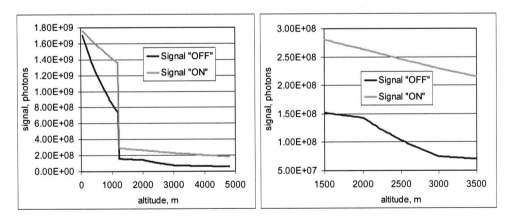

Figure 21. Illustration of "on" and "off" signals vertical profiles in water vapour DIAL – calculated signal for wavelength in the range of 815nm. The altitude distribution of the water vapours is not taken from measurement. Left: The profiles in the PBL and lower troposphere; Right: Zoom around the layer at higher altitude (see the text).

As we see from equations (105, 106), to obtain the concentration of the measured constituent with a DIAL we do not need instrumental constants, i.e., the DIAL method is "self-calibrated". The sensitivity of the method is determined by $\sigma_{on}^X - \sigma_{off}^X$. If we select a line with high σ_{on}^{X}, then the sensitivity will be high, but the radiation at λ_{on} will be absorbed at a shorter range. That is, in each practical case the selection of λ_{on} and λ_{off} shall be optimised with respect to both required sensitivity and range.

The DIAL measurement provides the number density of the constituent but not the mixing ratio, as it is in the Raman lidar. In case we need the mixing ratio, we have to determine in an independent way the air number density (from atmospheric models or from meteorological radiosondes).

One advantage of the DIAL method with respect to the alternative lidar technique (Raman) is the possibility to obtain the density profile of the measured constituent from the relatively strong elastic backscatter signal. This allows better resolution and daytime operation, as well as relatively compact lidar hardware. On the other hand, in the DIAL measurements the probing wavelength is attenuated in its propagation and this limits the range.

7.1.1. DIAL in the Near IR Spectral Range

A typical case when $\lambda_{off} \approx \lambda_{on}$ is in the near IR water vapour DIAL [*Bösenberg 2005*]. The water vapour absorption in this range exhibits bands containing a large number of narrow absorption lines (see in 4.5.1, Figure 13). Thus, "on" and "off" wavelengths may be close to each other, with a displacement varying from parts of nm to a few nm. The advantage of such close separation is the cancellation of the aerosol backscatter and extinction influence, as we have seen in equations (105, 106). On the other hand, the realisation of such DIAL is a spectroscopic and technical challenge.

The absorption cross-section and the shape of the line depend on temperature and pressure. That is, with the altitude variation of the atmospheric temperature and pressure the absorption coefficient at "on" wavelength will vary as well. Another problem is that the multitude of other lines close to the selected one with λ_{on}, combined with the temperature and pressure broadening of these lines, makes σ_{off}^{H2O} not negligible and also altitude dependent.

In such DIAL it is necessary to use lasers transmitting narrow lines and precisely tuned to "on" and "off" wavelength with respect to the molecular absorption line. The wavelength tuning, control and stability are critical issues. Fluctuations of the laser wavelength around the narrow absorption line result in unacceptable errors. Another critical issue is the spectral purity. A quite common phenomenon in the tunable lasers typically used in DIAL, is the so-called "amplified spontaneous emission". This is a radiation having the same direction as the laser beam, having much lower power, but outside the spectrum of the narrow-band laser line. In this way it is also outside of the molecular absorption line. As a result, this part of the radiation will propagate to and from the probed volume of the atmosphere without absorption, but still considered as "on" wavelength. Including such out-of-line "on" backscatter signal in equations (105, 106) will effectively "decrease" σ_{on}^{H2O} resulting in measurement error.

This combination of spectroscopic factors and the realistic laser line specifications determines a number of errors in the water vapour DIAL measurements, analysed in

[*Bösenberg 1998, Wulfmeyer and Bösenberg 1999*]. To the previously mentioned factors requiring corrections, we have to include the altitude profiles of the temperature and pressure, modifying the backscatter spectra by the Doppler broadening of the Rayleigh scattering [*Ansmann 1985, Ansmann and Bösenberg 1987*]. Requirements for the laser in water vapour DIAL are summarized in [*Wulfmayer 1998, Ertel 2004*].

During the years of DIAL development, there have been used a number of tunable laser types, including dye-lasers [*Browell et al. 1979*], Alexandrite lasers [*Bruneau et al. 1991, Wulfmayer et al. 1995*], Ti: Sapphire laser [*Ismail and Browell 1994, Ertel et al. 2005*] and optical parametric oscillators [*Ehret et al. 1998*].

DIAL measurements, including results from field campaigns are reported in [*Wulfmeyer 1999, Wulfmeyer 1999a, Ertel et al. 2001,* Bösenberg and Linné 2002, *Ertel 2004*]. Typically, the DIAL water vapour measurements are fast (i.e., a short integration time is necessary to reach convenient SNR) and may be done during the daytime. These allow the investigation of the dynamics of the PBL and mixing processes. The technical development and accumulated experience allow considerations for advanced DIAL instruments [*Wulfmeyer and Walther 2001, Wulfmeyer and Walther 2001a*], as well as DIAL network [*Wulfmeyer et al. 2003*].

It shall be mentioned here another application of elastic DIAL in the near IR. This is the determination of the atmospheric temperature using the temperature dependence in the absorption lines of the oxygen in the range 720nm-730nm [*Schwemmer et al. 1987, Theopold and Bösenberg 1993, Bösenberg 2005*].

A prospective application of near IR DIAL is the CO_2 concentration measurement, motivated by the "green-house" potential of this gas [*Ismail et al 2004, Nagai et al 2006*].

7.1.2. DIAL in the UV-Visible Spectral Range

The case when $\lambda_{off} \neq \lambda_{on}$ is typically the case of DIAL measurements of ozone in the UV. As we see in Figure 14, the wavelength dependence of the ozone absorption cross-section in the UV does not show prominent peaks. Due to this, the probing wavelengths need to be well separated to achieve sufficient $\sigma_{on}^{O3} - \sigma_{off}^{O3}$. With such separation the aerosol correction in Equation (104) is a basic problem, particularly, as mentioned above, at altitudes with gradients of the aerosol distribution [*Browell et al. 1985, Browell 1989*]. One way to solve this problem is the use of a third wavelength only for aerosol monitoring. This wavelength shall be close to λ_{off} and λ_{on}, so the values for the aerosol extinction and backscatter determined with it may be still extrapolated. At the same time the absorption cross section at this wavelength shall be negligible. To reduce the error due to the aerosol, a dual-DIAL approach also has been analysed [*Wang et al. 1996*].

The typical targets for UV DIAL measurements in the PBL and the lower troposphere are ozone, SO_2 and NO_2, as well as other industrial pollutions [*Schoulepnikoff et al. 1998*]. The SO_2 and NO_2 absorption (Figure 14), unlike for ozone, show structures containing sharp peaks. This allows the possibility for DIAL measurements of their concentrations with "on" and "off" wavelengths at closer separations than for ozone. From Figure 14 we may also see another problem with the UV DIAL: the absorption spectra of ozone, SO_2 and nitrogen oxides overlap in the UV. Also, these gases are typically present together in polluted PBL and lower troposphere, where the ratio of their concentration is also quite variable. This determines another problem in DIAL measurements in the UV, namely the cross talk of their absorption spectra in the "on" and "off" signals, which requires their simultaneous measurements as well. Anyway, this problem may be diminished by a proper selection of probing wavelengths. Such selection has been discussed in a number of works [*Maeda et al. 1990, Goers 1995, Proffitt and Langford 1997*].

There are several approaches to produce the radiation with λ_{off} and λ_{on} in UV DIAL for ozone measurements in the PBL and troposphere. One is based on tunable dye-lasers and subsequently producing second a harmonic [*Staehr et al. 1985, Edner et al. 1987, Fredriksson et al. 1981, Goers 1995, Fiorani 1996*]. Such transmitters are quite complicated but on the other hand they may produce optimal combinations of probing wavelengths. By using different combinations of "on" and "off" wavelengths, the probing wavelength pairs may be optimised for detection of a particular gas (i.e., ozone, SO_2, NO_2) in a way to minimise the influence of the absorption spectra from the other gases. Another tunable laser used in the UV DIAL is the Optical Parametric Oscillator (OPO) [*Fix et al. 2002*].

The alternative approach makes use of fixed wavelengths, produced by stimulated Raman shifting. The stimulated Raman shifting takes place in H_2 or D_2, or CH_4, mixed with buffer gases. The pumping sources are either KrF eximer laser or the fourth harmonic of the Nd:YAG laser. This process produces one or two probing wavelengths for each gas, determined by gas molecular constants [*Ziao et al. 1992, Kempfer et al. 1994, Schoulepnikoff et al. 1997, Simeonov et al. 1998*]. The advantage of such a transmitter is in the compact set-up, easy maintenance and higher efficiency. On the other hand, it is optimal for ozone DIAL but not necessarily for other gases.

The UV DIAL has successfully operating for more than two decades [*Browell 1982*]. DIAL systems are often assembled in a van or a truck, in this way ensuring mobility in investigation of industrial pollutions. Such DIAL based on tunable dye laser and adaptable for measurement of different gases with different probing wavelengths was previously quoted [*Edner et al. 1987, Fredriksson et al. 1981*]. Measurements of SO_2, NO_2 and Cl_2 with this system are reported in [*Egeback et a.l 1984, Edner et al. 1987a, Edner et al. 1989*]. The detection of nitrogen oxides was reported in [*Fredriksson and Hertz 1984, Kölsch et al. 1989*]. A typical application of such mobile UV DIALs is the participation in complex field campaigns for pollution measurements [*Fiorani 1996, Durieux et al. 1998*].

The use of a novelty shot-per-shot signal acquisition system allowed a more precise evaluation of the correction factors in UV DIAL for ozone [*Fiorani 1996, Fiorani et al. 1997*]. A very good comparison is found between the DIAL ozone measurements in the PBL with the measurements of sensors on balloon and on motoglider (Figures 22 and 23). Such a fast signal acquisition technique in DIAL may be also combined with methods for wind velocity measurement [*Fiorani et al. 1998*].

The practical perspective of ozone DIAL using stimulated Raman shifters to produce the probing wavelengths is demonstrated in [*Ancellet et al. 1989*] and analysed further in [*Papayannis et al. 1990*]. Another description of such ozone DIAL may be found in the already quoted [*Kempfer et al. 1994*] where results from systematic observations are presented in [*Carnuth et al. 2002*]. The validity of such DIAL measurements was also proved by comparison with different types of ozone radiosondes [*Beekmann et al. 1994*]. A number of groups in Europe developed successfully operating ozone DIALs, implementing procedures for the aerosol correction. This motivated an intercomparison campaign between these lidars [*Bösenberg et al. 1993*].

The UV DIAL is applied also for measurements of the stratospheric ozone, motivated by the global ozone problem. The potential of such a system was first evaluated with dye-lasers [*Mégie et al. 1977*]. A number of stratospheric ozone DIAL systems are presently operational around the world, based on Raman shifters pumped by excimer lasers. A representative example for such system is given in [*Stefanutti et al. 1992a*].

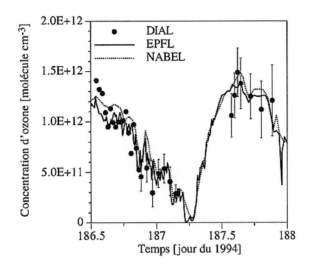

Figure 22. Intercomparison between ozone measurements performed by two in situ analysers (EPFL, NABEL) and with the DIAL directed horizontally. The time is in day 1994: as example, 187.5 corresponds to noon (solar time) of the 187th day of 1994. The vertical bars indicate the statistical error. Payerne, 5-6 July 1994; 1000 (left) and 2000 (right) laser shots; from [Fiorani 1996]; Reproduction is with the permission and courtesy of Dr. Luca Fiorani.

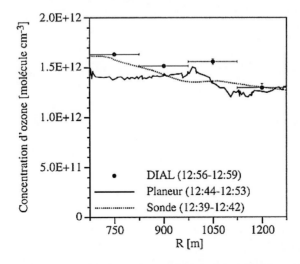

Figure 23. Intercomparison between ozone profiles measured by the DIAL directed vertically, motoglider (Planeur) and balloon (Sonde). Payerne, 30 June 1994; 5000 laser shots; from [Fiorani 1996]. Reproduction is with the permission and courtesy of Dr. Luca Fiorani.

7.2. Raman-DIAL

The Raman DIAL transmits one probing wavelength and receives two Raman shifted backscatter signals, respectively from oxygen and nitrogen, i.e., atmospheric constituents with high abundance and constant mixing ratio. When the Raman shifted wavelengths are into the absorption band of atmospheric constituent with concentration $N_X(r)$, Equation (84) may be written for each signal in the following way:

$$E_{N2}(r) = E_L K_{N2} \eta_{N2} O_{N2}(r) A \frac{c\tau}{r^2} \beta_{N2}^{Ram}(r) exp\left\{-\int_0^r \{\alpha_L(r') + \alpha_{N2}(r') + \sigma_L N_X(r) + \sigma_{N2} N_X(r)\}dr'\right\} \quad (107)$$

$$E_{O2}(r) = E_L K_{O2} \eta_{O2} O_{O2}(r) A \frac{c\tau}{r^2} \beta_{O2}^{Ram}(r) exp\left\{-\int_0^r \{\alpha_L(r') + \alpha_{O2}(r') + \sigma_L N_X(r) + \sigma_{O2} N_X(r)\}dr'\right\} \quad (107a)$$

After taking the ratio of equations (107, 107a), a similar transformation as with the ratio of equations (102-102a) and after passing to discreet range resolution, we present the following expression for $N_X(r)$

$$N_X(r) = (\sigma_{O2} - \sigma_{N2})^{-1}\left\{\frac{1}{\Delta r}\left[ln\frac{S_{N2}(r+\Delta r)S_{O2}(r)}{S_{N2}(r)S_{O2}(r+\Delta r)}\right] + \left[\alpha_{N2}^{mol}(r) + \alpha_{N2}^{aer}(r) - \alpha_{O2}^{mol}(r) - \alpha_{O2}^{aer}(r)\right]\right\} \quad (108)$$

From Equation (108), we see that there is no term containing the log-derivative of the aerosol backscatter coefficients, as is the case in the backscatter DIAL. In this way, the uncertainty due to the aerosol backscatter profile is cancelled, which is one advantage of this method. The difference $\alpha_{N2}^{mol}(r) - \alpha_{O2}^{mol}(r)$ may be evaluated from models or radiosondes measurements. Indeed, a correction due to the difference $\alpha_{N2}^{aer}(r) - \alpha_{O2}^{aer}(r)$ shall be considered.

A disadvantage of the Raman DIAL compared to the backscatter DIAL is the lower level of the backscatter signal. This makes it more convenient for closer ranges and at lower altitudes, i.e., PBL and lower troposphere. There, the advantage and the disadvantage are quite balanced, since at those altitudes, the problem with the aerosol correction in the backscatter DIAL is most pronounced, due to the large gradients in the aerosol vertical distribution.

The idea for the Raman DIAL method was presented with respect to the ozone absorption correction in water vapour measurements in the UV spectral range below 300nm [*Renault et al. 1980, Renault and Capitini 1988*]. The idea was analysed further in [*Cooney 1986, Cooney 1987*]. To avoid the interference with the aerosol extinction terms in Equation (107), a modification based on the rotational and rotational-vibrational Raman spectrum is discussed in [*Reichardt et al. 2000*].

A representative example of Raman DIAL measurement of ozone in the lower atmosphere, combined with Raman lidar measurement of water vapour, is presented in [*Lazzarotto et al. 2001*]. Notwithstanding the lower level of the Raman signal, it was applied also in stratospheric ozone measurements during the period of severe volcanic eruption leading to an increase in the aerosol content at those altitudes [*Mc Gee et al. 1993*].

8. AIRBORNE AND SPACEBORNE LIDARS

8.1. Airborne Lidars: Advantages

The previous chapters describe the basic principles of the atmospheric lidars. The examples were predominantly for lidars operating from the surface and probing the atmosphere in an upward direction. When the lidar operates continuously from a fixed site (or slowly moving [*Ertel et al. 2001*]), the measurements present the temporal development of the investigated atmospheric parameters above it.

The airborne atmospheric lidars apply indeed the same basic principles of laser beam propagation, scattering and absorption as the surface based lidars. The differences, the advantages and the problems come from the specifics of the airborne platform.

The air mass velocity is much lower compared to the aircraft cruising speed. This allows consideration of the atmosphere "frozen" during the time needed for the aircraft to cover a selected region. In this way the lidar resolves the variation of the atmospheric structures along the flight track, i.e., the spatial variation of the atmospheric parameters and structure.

The airborne lidar may operate from a platform above the atmospheric layers with high attenuation or opacity (fog, clouds). In this way it realises two advantages. One is that the lidar is not weather dependent. The second is that it allows investigations that are not possible in principle from the surface, such as processes taking place above the top of a cloud.

The airborne lidar may investigate the atmosphere above regions not accessible in another way or accessible, but at prohibitive costs, due to the lack of infrastructure. Examples are the fast deployments and measurements over oceans, in the Polar Regions, above tropical forests, etc.

For surface-based lidars, both the signal dependence r^{-2} from Equation (12) and the exponential altitude dependence of the molecular density from equations (34, 35) decrease the backscatter signal. This results in the strong requirement for a high dynamic range in the signal detection system. The airborne lidar may be up-looking and down-looking. For up-looking lidar there is the same requirement for a high dynamic range. For the down-looking airborne lidar, the exponential altitude dependence of the molecular density tends to balance the dependence r^{-2} from Equation (12), i.e., there is a less strong requirement for a high dynamic range of the detector. Indeed, this assumes that the magnitude of the backscatter return from the surface is not of interest.

On the other hand, the airborne installation of an atmospheric lidar requires the solution of several problems not needed for a surface-based one. One problem is the availability of the aircraft and its flight time. Another is the limited volume and mass on board. As a result, the airborne lidar configurations shall be compact, i.e., having a receiver with a smaller size. Also, the lidar mechanical system shall resist shocks and vibrations, and the lidar electronics shall be capable to operate with the electrical supply from the aircraft and in its radio frequency environment. There are indeed solutions for all these problems, but they come with a considerable installation cost. A separate but also important issue is eye-safety for the down-looking airborne lidars. The beam divergence and wavelength shall be taken into account to determine the safe combinations of flight altitudes and transmitted powers [*ANSI 1993*] and at the same time to fulfil the measurement requirements.

Nevertheless, with all of the inherent limitations and problems, the first airborne lidar demonstrations took place in the late 1960s with the elastic backscatter type, as the most simple and mature for installation at the time. A detailed historical review of the airborne lidars is presented in [*McCormick 2005*]. Presently the most widespread airborne lidars are the elastic backscatter and the elastic DIAL.

8.2. Airborne Elastic Backscatter Lidars

A very convenient field of application for the airborne down-looking elastic backscatter lidar is the investigation of the PBL. One representative early example is the observation of convective cells and entrainment zone in marine PBL [*Melfi et al. 1985*]. As part of a larger instrument complex, the elastic airborne lidar measurements described the PBL structure during specific meteorological conditions above sea [*Dupont et al. 1994, Flamant and Pelon 1996, Flamant et al. 1997*]. Another advantage of the airborne lidar, the possibility to investigate the cloud-top, is demonstrated in [*Spinhirne et al. 1982, Spinhirne 1983*].

Over complex topography and urban areas the PBL develops horizontally variable structures containing layers of pollutions and aerosol. The down-looking lidar reveals the horizontal expanse of such structures. Observation of aerosol pollution layers in the PBL above urban agglomeration, their accumulation and transport due to the complex terrain of the surroundings, is reported in [*Wakimoto and McElroy 1986, McElroy and Smith 1986*]. The emission and the distribution of the aerosols in a valley are studied by lidar in [*Hoff et al. 1997*]. In the PBL above mountain slopes the venting during convective development may elevate aerosol layers to higher altitudes. This effect is observed in [*Nyeki et al. 2000*] by backscatter lidar on an aircraft flying above a high Alpine ridge.

Aerosol measurements above the ocean, including observations of Sahara dust transport are also carried out with down-looking elastic airborne lidar [*Flamant et al. 2000, Chazette et al. 2001, Pelon et al. 2002*].

The need for better resolution and faster data analysis with the elastic backscatter airborne lidars motivated development of scanning [*Palm et al. 1994*] and calibration procedures [*Spinhirne et al. 1997*]. The contribution of the surface return in the down-looking lidar signal processing is analysed in [*Hooper and Gerber 1986*].

As the lidar is on aircraft, it is possible to measure along a selected track. In this way it is easy to collocate the measurements from the lidar and from satellite-borne sensors. Thus, the airborne lidars contributed substantially to the validation of the passive satellite-borne optical sensor for the aerosol measurements of SAM II [*Russell et al. 1981*]. Going forward, it shall be noted that the measurements of the first space-borne lidar mission (it will be described further) find an excellent validation also with down-looking airborne lidar measurements [*Renger et al. 1997, Schwemmer et al. 1997*].

A substantial contribution of the airborne elastic backscatter lidars is the study of the stratospheric aerosol layers after volcanic eruptions [*McCormick et al. 1983, McCormick et al. 1983a, Wirth 1994, McCormick et al. 1995*]. The airborne lidar measurements in the polar region made possible the study and the classification of the PSCs [*Poole and McCormick*

1988, Browell et al. 1990, Toon et al. 1990]. The observations by airborne elastic backscatter lidar contributed also to the understanding of one principle mechanism for PSC formation, the Lee-waves [*Godin et al. 1994, Wirth et al. 1999, Dörnbrack et al. 2002*].

8.3. Airborne DIAL

The first airborne DIAL operation was for measurements of ozone and water vapour [*Browell et al. 1979, Browell et al. 1983*]. The lidar transmitter was based on a dye-laser, pumped by Nd: YAG harmonics. Both ozone and water vapour DIALs use the residual pumping laser radiation for elastic backscatter measurements, i.e., they combine DIAL ozone or water vapour measurements with aerosol profiling. This experiment marked the start of a long-term airborne DIAL activity at the NASA Langley Research Center, consisting of both successive upgrades of the lidar set-up and numerous field campaigns [*Browell et al. 1993, Newell et al. 1996, Browell et al. 2001, Browell et al. 2003*]. The campaign measurements typically combine the altitude profiles of ozone or water vapour, and aerosol, as well as the meteorological parameters from auxiliary sensors, and in some cases, satellite observations. During the successive campaigns a large and reliable database was accumulated, providing the possibility to validate models for atmospheric ozone [*Douglass et al. 2001*].

Long-term airborne DIAL development and participation in field campaigns have been carried in DLR. The first realisations were based on dye-laser [*Ehret et al. 1993*] and recently on all-solid-state OPO [*Ehret et al. 1998, Poberaj et al. 2002*]. The measurement included investigation of PBL dynamics, in which the tracer is not only aerosol but water vapour [*Kiemle et al. 1995, Kiemle et al. 1997*]. The measurement campaigns with water vapour DIAL took place in various geographic regions offered the possibility to contribute to its climatology [*Ehret et al. 1999, Ehret et al. 2006*]. Airborne ozone DIAL is also realised and applied for study in the Arctic atmosphere [*Wirth and Renger 1996*].

It shall be noted that a number of the instrument and algorithm developments for DIALs were carried out in relation to the airborne DIAL programs. Examples are the development of novelty water vapour transmitter [*Bruneau et al. 1991, Bruneau et al. 2001*] and its application [*Flamant et al. 2004*].

An interesting airborne lidar experiment is presented in [*Uthe 1991*]. The lidar is DIAL using a CO2 laser at 10.6µm but also with additional elastic backscatter measurements at 1064nm and 532nm, and with a channel to detect also the fluorescence from artificial aerosol-tracers at 600nm. The tracer is released into the atmosphere to "visualise" the probable path for distribution of pollutions.

8.4. Compact Elastic Backscatter Lidars on a High-Altitude Aircraft

Since 1996 the high-altitude research aircraft M-55 "Geophysica" has been used by a team of European institutions for studies in the Upper Troposphere and the Lower Stratosphere (see Figure 24). The aircraft accommodates a number of *in situ* and remote sensing instruments [*Stefanutti et al. 1999*]. With various options of its research payload, this aircraft participated in a number of field campaigns: APE-POLECAT, APE-GAIA, APE-THESEO, EUPLEX, TROCCINOX, SCOUT-O3 and ENVISAT Validation.

Two compact elastic backscatter lidars were realised and installed [*Mitev et al. 2002*]. The lidars were developed and optimised for elastic backscatter and polarisation measurements of aerosol and clouds close to the aircraft (to a few km). The probing wavelength is the second harmonic of a micro-pulse Nd:YAG laser. The lidars are referred to as MAL 1 and MAL 2, abbreviation of "Miniature aerosol lidar Mk 1" and Mk 2, respectively. On the aircraft one is installed to probe in upward direction and the other probes in a downward direction. The lidars are compact, each having total mass of 27kg, including the pressurised environment protection vessel. After being installed, they operate in an unattended way. Figure 25 shows the view of the two lidars in their pressurised vessels, ready for installation. The specifications of the lidar subsystems are given in Table 3.

Figure 24. High-altitude research aircraft M-55 "Geophysica."

Table 3. Specifications of the subsystems of the lidars MAL

Laser type	Nd-YAG micro-chip, passive Q-switched, various suppliers
Wavelength/average power	532 nm /16-20mW
Laser pulse rep. rate	5 KHz - 6 kHz
Polarisation	linear
Transmitted beam divergence	0.25 mrad full angle
Receiver type and aperture	refractor, 50 mm $\varnothing$
Receiver field of view	0.5 mrad full angle
Detectors	PMT photon counting module HAMAMATSU H6240-01
Dark noise	35 – 160 cps
PMT output pulses	TTL positive, 30 ns width
Count rate sensitivity	typ. 3×10^5 cps/pW
Acquisition range gate	Minimum value 10 m
Number of range gates	Variable, from 1024
Single measurement, duration	about 3.5s
Dimensions of the lidar	MAL 1: 510x320x170mm^3 /MAL 2: 440x320x170mm^3
Mass of the lidar	27 kg in the pressurised vessel
Consumed current / voltage	max. 3A/28V DC, including the temperature control
Operation control	continuous housekeeping parameter record
Safety loops in case of:	de-pressurization, overheating, overcooling

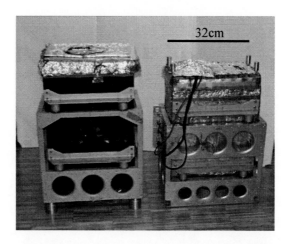

Figure 25. Left: Backscatter lidar MAL1, refered also as "MAL-up", for "upward-probing"; Right: Backscatter lidar MAL2, refered also as "MAL-down", for "downward-probing". The lidars are shown in their pressurised containers, ready for installation on M55 "Geophysica".

The targets for these lidars, in missions studying the upper troposphere and lower stratosphere are the PSCs, Ultrathin Tropical Tropopause Clouds (UTTC) and cirrus clouds. Both PSCs and UTTCs are connected to the chemical composition of the atmosphere, and they affect this chemical composition after being formed. The detection of such clouds allows the connection of the atmospheric composition detected by *in situ* instruments on the aircraft with the aerosol and cloud backscatter and depolarisation coefficients.

During APE-THESEO, one of the lidars participated in a cirrus cloud study together with a number of *in situ* instruments [*Thomas et al. 2002*]. A spectacular PSC observed by this lidar in a Lee-wave above the Antarctic Peninsula is described in [*Cairo et al. 2004*]. The participation in the ENVISAT validation campaign provided the opportunity for a correlation of the MAL1 and MAL2 measurements with another airborne lidar [*Fiocco et al. 1999, Matthey et al. 2003*].

Examples of measurements performed with the described lidars are presented in Figures 26-29. Figure 26 shows a cirrus cloud cross-section, combining the simultaneous measurements of the two lidars when M-55 crossed this cloud, displaying the complicated structure of such clouds.

During a campaign flight on January 15, 2003, in the EC project EUPLEX (*Europan Polar stratospheric cloud and Lee-wave Experiment* - http://www.nilu.no/euplex/), the lidar MAL1 detected a PSC at 22-24km, presented in Figure 27. The figure shows also the terrain profile below the aircraft. We may see that the cloud is detected above a higher part of a mountain area. A high value of the depolarisation ratio was observed (between 30%-45%) consistent with Lee-waves PSC type II [*Toon et al. 1990*].

One of the objectives of the EC project TROCCINOX (*Tropical Convection, Cirrus and Nitrogen Oxides Experiment* - http://www.pa.op.dlr.de/troccinox/) was aerosol and cirrus measurement in the upper tropical troposphere. The campaign took place from Aracatuba, Brazil and was performed with two research aircrafts: M-55 "Geophysica" and DLR Falcon, the latter equipped with water vapour DIAL. The transfer flights of the two aircrafts from Europe to Brazil over the Atlantic provided the opportunuty for a simultaneous observation of the UTTC above the Atlantic [*Martucci et al. 2006*]. Figures 28 and 29 present the measurements performed by MAL2, showing the presence of ultra-thin clouds at an altitude of 17 km, a convective cloud above the South America coast and clouds at low altitudes above the ocean.

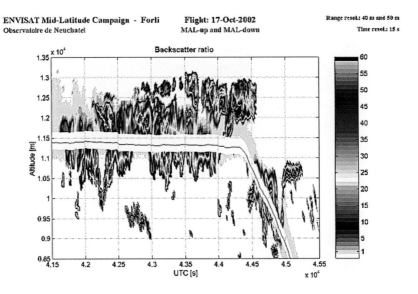

Figure 26. Backscatter ratio of a cirrus cloud; flight on 17 October 2002, Forli-Italy. The black line shows the altitude of M55. The aircraft crosses the cloud. The figure shows the data both from MAL1 and MAL2. The grey scale on the right shows the backscatter ratio.

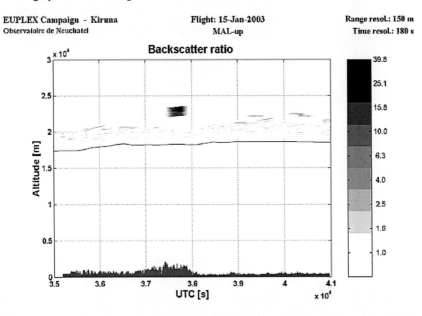

Figure 27. Lee-wave PSC observed during the campaign EUPLEX flight on January 15, 2003 (Kiruna, Sweden). The backscatter ratio is given in the grey scale on the right. The cross-section of the surface along the flight path is added to show that the cloud appears above the highest part of the terrain. The black line shows the flight altitude of M55 "Geophysica".

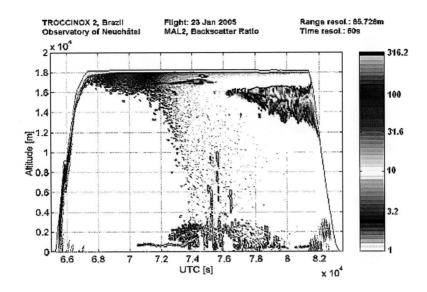

Figure 28. MAL2 observations during TROCCINOX flight on 23 January 2005. Backscatter ratio: gray scale on the right. Subvisible clouds appear between 72,000 and74,000 s and 16-17km. [Martucci et al. 2006].

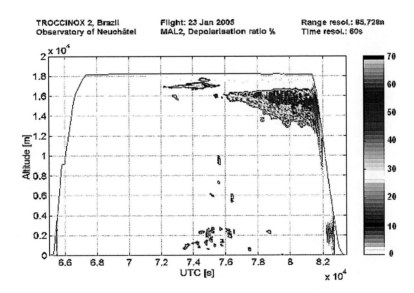

Figure 29. MAL2 observations during TROCCINOX flight on 23 January 2005. Depolarisation ratio: gray scale. As iin Fig. 28, subvisible clouds appear between 72,000 and74,000 s and 16-17km. [Martucci et al 2006].

8.5. Spaceborne Lidars

The advantages offered by an orbiting lidar for Earth observation are clear: The lidar measurements will enable a global coverage, particularly when the lidar is on a polar orbit. This will include measurements over critical regions presently not routinely accessible. Unlike the passive optical satellite-borne sensors for aerosol, water vapour, or temperature, etc., the lidar profile will be obtained without *a priori* assumptions for the altitude distribution of the measured parameter, as well as without constraints of the position of natural light sources.

The technological challenges to install the lidar in orbit are also clear: The limitations of mass and the volume of the lidar, and electrical power and communication bandwidth, are more stringent than for the airborne lidars. A new requirement is reliability of the hardware and its capability to operate without attendance and maintenance for the duration of the space mission (sometimes years). Another requirement is resistance to the environmental conditions in space: large temperature changes, vacuum, and radiation. It is also imperative that the lidar hardware shall withstand the shocks and vibrations during launch.

Motivated by the advantages, the space agencies stated to consider orbiting lidar missions. A number of missions were studied and a very informative historical overview on this subject is presented by one of the pioneers in this field [*McCormick 2005*]. The lidar community supported these considerations with numerical analysis of the necessary performance for the various types of lidars [*Russell et al. 1982, Russell and Morley 1982, Uchino et al. 1986, Curran 1989, Ismail and Browell 1989, Browell et al. 1998, Liu et al. 2000*]. Also, it started a development of algorithms and procedures for advanced satellite-borne lidars data evaluation [*Doutriaux-Boucher 1998, Balin et al. 1999*].

The first space lidar mission was NASA LITE (Lidar In-Space Technology Experiment). LITE is a three-wavelength backscatter lidar. This lidar mission was only for demonstration of the feasibility and the potential of lidar in space. The transmitter was a flash-lamp pumped Nd:YAG laser with second and third harmonics, and 1m diameter of the receiving telescope. The scientific planning of this space lidar mission is described in [*McCormick et al. 1993*]. The lidar was assembled in a pressurised container and was taken to orbit by the Space Shuttle Discovery, mission STS-64. The launch took place on September 9, 1994. the landing was on September 20, 1994. An overview of the mission is presented in [*McCormick 2005*].

A number of validation measurements of the LITE lidar were organised. Two of those are mentioned above [*Renger et al. 1997, Schwemmer et al. 1997*]. Another correlative measurement by airborne lidar is presented in [*Strawbridge and Hoff 1996*]. A validation case by a ground-based lidar is presented in [*Cuomo et al. 1998*].

In demonstrating the potential of the lidar-on-orbit for Earth observation, LITE was a clear success. The observations collected during its operation described a large number of phenomena. An investigation of ultra thin cirrus clouds with large horizontal expanse at altitudes around the tropical tropopause is described in [*Winker and Trepte 1998, Omar and Gardner 2001*]. The LITE data are used together with Meteosat observations and atmospheric model to investigate the Sahara dust transport [*Karyampudi et al. 1999*]. The stratospheric aerosol observation is presented in [*Osborn et al. 1998*] and tropospheric aerosol layers in [*Kent et al. 1998*]. LITE also observed transport of urban and industrial pollution aerosol [*Hoff and Strawbridge 1997*] as well as debris from biomass burning [*Grant et al. 1997*]. The

LITE measurements also demonstrated the possibility to determine the surface wind velocity from the surface return in space lidars [*Menzies et al. 1998*].

The NASA GLAS (Geoscience Laser Altimeter) on the ICESat (Ice, Cloud and land Elevation Satellite) is the first long-term mission of Earth observation lidar [*Zwally et al. 2002*]. The lidar is elastic backscatter, operating with a transmitter consisting of three independently operating diode lasers pumped Nd:YAG lasers and their second harmonic. This ensures redundancy of the laser operation. The expected lifetime of the mission is determined by the laser lifetime and is estimated to be two years. GLAS was launched on 13 January 2003. Its primary objectives are the measurements of the ice-shield variation, but it also may detect clouds and aerosols in the atmosphere [*Schutz et al. 2005, Abshire et al. 2005*]. First results from the lidar check and calibration are presented in [*Palm et al. 2004*] and from cloud observation in [*Dessler et al. 2006*].

ALISSA was a joint French-Russian elastic backscatter polarisation lidar in the module "Priroda" on the Space Station "Mir". The lidar operated with four independent Nd:YAG lasers. After its installation encouraging preliminary results from the system verification were presented [*Hauchecorne et al. 1998*]. The intention was for a long-term operation, where the astronauts were also supposed to participate in the lidar maintenance. Anyway, the termination of the "Mir" Station put a premature end to this lidar.

CALIPSO is a joint NASA-CNES satellite on a polar orbit. It carries a dual-wavelength polarisation elastic backscatter lidar CALIOP (Cloud-Aerosol Lidar with Orthogonal Polarization). The lidar operates with Nd:YAG laser and its second harmonic (i.e., 1064nm and 532nm) has a pulse energy of 110mJ per each wavelength. The pulse repetition rate of the laser is 20-25Hz. The lidar receiver has a diameter of 1m. A detailed description of the lidar may be found in [*Winker et al. 2004*]. The launch was on 28 April 2006 with a Delta II rocket. A lifetime of three years is expected. An overview of the first months of CALIPSO operation is presented in [*Winker et al. 2006*]. Preliminary results may be also seen at webpages: http://www-calipso.larc.nasa.gov/,   http://www-calipso.larc.nasa.gov/products /lidar/, and http://www.icare.univ-lille1.fr/calipso/browse. Based on the first results, the lidar community has good reason to expect that this space lidar will be a very successful mission.

Figure 30 depicts one such result (http://www-calipso.larc.nasa.gov/products/lidar/). The measurement is performed above Antarctica and likely presents a PSC at altitudes to 22-23km. A lower cloud structure is also presented. It is noted in the description of the first results that the signal is still high at ranges (altitudes) below the surface, which is an artifact following from the afterpulse effects in the photomultipliers, as mentioned in 2.4.4.

For several years, ESA studied a spaceborne water vapour DIAL WALES (WAter vapour Lidar Experiment in Space), operating in the near IR range. A number of studies concerning the requirements and the processing algorithm were published [*Bauer et al. 2004, Summa et al. 2004*], as well as realisation of demonstrators for some of its critical systems [*Matthey et al. 2006*]. Anyway, due to various reasons, including demanding hardware requirements, the further development of this mission is pending.

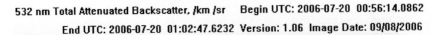

532 nm Total Attenuated Backscatter, /km /sr   Begin UTC: 2006-07-20 00:56:14.0862
End UTC: 2006-07-20 01:02:47.6232 Version: 1.06 Image Date: 09/08/2006

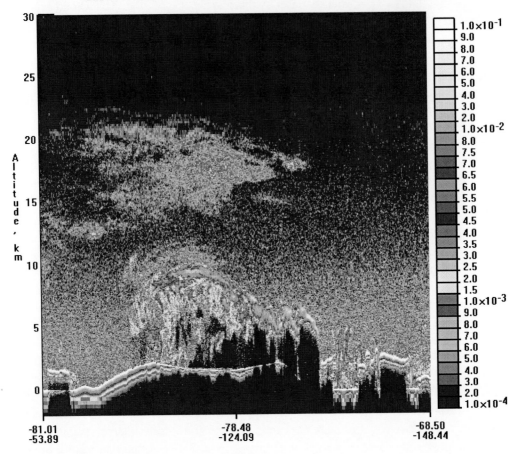

Figure 30. An example of measurement performed by space lidar CALIOP on CALIPSO, above Antarctica on 20 July 2006. The vertical axis presents the altitude; the upper and the lower horizontal scales present respectively the latitude and the longitude; image from "http://www-calipso.larc.nasa.gov/products/lidar/". Reproduction is with the permission and courtesy of NASA and IPSL; with acknowledgements to Dr. Jacques Pelon for his attention and helpful suggestions.

Another space lidar in preparation in ESA is ALADIN (Atmospheric Lidar Doppler Instrument) for the Atmospheric Dynamics Mission ADM-Aeolus – see the artistic impression in Figure 31 (from http://www.esa.int/esaLP/ESAVO62VMOC_LPadmaeolus_0. html#subhead4). The specifications of the critical lidar subsystems are given in Table 4 (from the same web-page). The preparation of this mission is in progress [*Dubock et al. 2006*], with an airborne demonstrator of the lidar presented in [*Reitebuch et al. 2004*]. This lidar will use the incoherent Doppler principle for wind measurements, mentioned in Chapter 5.8. The lidar ALADIN will also be capable of determining the aerosol altitude profile [*Ansmann et al. 2006*].

ESA also considers a mission, EarthCARE, in which an elastic backscatter lidar will be also used – ATLID (Atmospheric backscatter Lidar). The concept for the mission and the lidar are presented in [*Hélière et a.l 2006*].

The elastic backscatter lidar is considered capable of withstanding years of space flight to other planets, as well as the shock of landing. A lidar for dust measurement in the atmosphere of Mars is in realisation for the PHOENIX [*Whiteway et al. 2006*].

Figure 31. Aeolus with laser beam pointing at the Earth's atmosphere. Reproduction is with the permission and courtesy of ESA.

Table 4. *Technical specifications of the ALADIN instrument (from*
http://www.esa.int/esaLP/ESAVO62VMOC_LPadmaeolus_0.html#subhead4)

Subsystem	Parameters	Value
Transmitter	Wavelength	355 nm
	Pulse energy	130 mJ (150 mJ goal)
	Repetition rate	100 Hz
	Line width	30 MHz
	Duty cycle	25%
Receiver	Fizeau line width (Mie)	30 MHz
	Double Fabry-Perot (Rayleigh): Line width/Spacing	2 GHz/5 GHz
	Optical efficiency (Mie/Rayleigh)	3.1% / 4.6%
	Detector quantum efficiency (Mie/Rayleigh)	75%

Table 4. (Continued)

Subsystem	Parameters	Value
Signal Processing	Altitude range (Mie + Rayleigh)	-1 to +26.5 km (extendable)
	Vertical resolution	1 km (adjustable)
	On-chip horizontal accumulation length	3.5 km (adjustable)
	Processing integration length	50 km
Opto-mechanical Subsystem	Telescope diameter	1.1 m
	Optical efficiency	0.8

Reproduction of the values is with the permission and courtesy of ESA.

9. CONCLUSION AND PERSPECTIVES

Presently the atmospheric lidar is a mature technique, based on reliable hardware set-up and developed processing algorithms. The lidar contributes to measurements of critical atmospheric parameters. Those include aerosol extinction and backscatter, clouds, temperature, the concentration of water vapour, ozone, industrial pollutants such as SO_2 and others. These measurements are performed in many groups and in a repeatable way over a number of years.

The achieved altitudes in lidar measurements are from a few hundred of meters above the surface to the upper stratosphere and mesosphere. The lidar potential for continuous measurements make it possible to investigate the variation of the atmospheric parameters from a few hours and days (the scales for meteorological events and diurnal cycles) to a number of years (the time scale required in climatology).

During the years of its development the atmospheric lidar already contributed extensively to the investigation of critical atmospheric phenomena. In the PBL, these are the diurnal cycles, fluxes and transport of aerosol, water vapour and pollution gases. In the troposphere, these are the long-range transport of desert dust and industrial pollutions, statistics of subvisible clouds. In the stratosphere, these are the stratospheric contamination after substantial volcanic eruptions, the stratospheric ozone changes, PSC classification and climatology, the stratospheric temperature including its short-term variations showing wave processes, and its long-term climatology.

The results from the measurements are also straightforward for interpretation and are in synergy with the measurements from the other instruments used in atmospheric and environmental studies.

The development of the lidars follows the progress of the photonics devices: powerful lasers, detectors and optics. There are convincing signs that this progress will continue, and that it will be efficiently implemented in the lidars. This will not only improve the lidar performances with more compact and stable lidar set-ups, but will also make possible lidar applications presently not considered. Also, lidar applications now demonstrated in just a few groups or in unique set-ups may become more widespread and even routine, as the Doppler wind lidar and the high-resolution lidar for extinction and backscatter.

We may expect that the recent trends in the applications of the atmospheric lidars will continue in the future.

Most of the measurements will still come from surface-based lidars. The lidar will be more and more used in synergy with other instruments for atmospheric measurements. One perspective is the establishment of large-scale lidar networks based on the surface-based lidars, with the previously-mentioned EARLINET being a leading example. Such networks will contribute to the climatology of critical atmospheric constituents as aerosol and water vapour, as well as to the tracing of pollution transport paths.

Another trend is the lidar application in field campaigns dedicated to the study of critical atmospheric and environmental phenomena. In such campaigns, the lidar will operate together with other instruments for atmospheric measurements and atmospheric models, contributing to the detailed description of the studied phenomena.

The further development and application of airborne lidars will be boosted by progress in photonic technology. The future compact and high-performance lidars may be used from smaller crafts, with fewer requirements for support, and following from these, to be used in more cases than presently.

The place of space-borne lidars will be to ensure real global coverage of atmospheric measurements, needed both for weather forecasting and to understand the controversial issues in climatic changes. Indeed, these lidars will always require unique and highest-performance realisation and support.

Each of these prospective trends has its place in atmospheric studies. The future of the lidar will be based on the combination and the synergy of all of them.

REFERENCES

Abshire J. B., Sun, X., Riris, H. et al. (2005). Geoscience Laser Altimeter System (GLAS) on the ICESat Mission: On-orbit measurement performance, *Geophys. Res. Lett.,* 32, L21S02, doi: 10.1029/2005GL024028.

Ackermann J. (1998). The Extinction-to-Backscatter Ratio of Tropospheric Aerosol: A Numerical Study, *J. Atmos. and Ocean. Technology*, 15, 1043-1050.

Althausen D., Müller, D., Ansmann, A., et al. (2000). Scanning 6-Wavelength 11-Channel Aerosol Lidar, *J. Atmos. and Ocean. Technology*, 17, 1469-1482.

d'Almeida G.A., Koepke, P. & Shettle, E.P. (1991). *Atmospheric Aerosols: Global Climatology and Radiative Characteristics, A. Deepak Publishers.*

Ancellet G., Papayanis, A., Pelon, J., et al. (1989). DIAL tropospheric ozone measurement using Nd. YAG laser and the Raman shifting technique, *J. Atmos. and Ocean. Technology.*, 6, 832-839.

Anderson G.P., Clough, S.A., Kneizys, F.X. et al. (1986). *Atmospheric Constituents Profiles (0-120 km), AFGL-TR-86-0110.*

ANSI. American National Standard for the Safe Use of Lasers, American National Standards Institute, 1993.

Ansmann A., (1985). Errors in ground-based water-vapor DIAL measurements due to Doppler-broadened Rayleigh backscattering, *Appl. Opt.*, 24, 3476-3480.

Ansmann A. & Bösenberg, J. (1987). Correction scheme for spectral broadening by Rayleigh scattering in differential absorption lidar measurements of water vapour in the troposphere, *Appl. Opt.*, 26, 3026-3032.

Ansmann A., Riebesell, M. & Weitkamp, C. (1990). Measurement of atmospheric aerosol extinction profiles with a Raman lidar, *Opt. Lett.*, 15, 746-748.

Ansmann A., Wandinger, U., Riebesell, M., et al. (1992). Independent measurement of extinction and backscatter profiles in cirrus clouds using a combined Raman elastic-backscatter Lidar, *Appl. Opt.*, 31, 7113-7131.

Ansmann A., Bösenberg, J., Chaikovsky, A., et al. (2003). Long-range transport of Sahara dust to northern Europe: The 11-16 October 2001 outbreak observed with EARLINET, *J. Geophys. Res.*, 108 (D24), 4783-4798.

Ansmann A. & Müller, D. (2005). Lidar in Atmospheric Aerosol Particles, p.p. 105-141, *in Weitkamp C., (Editor), Lidar: Range-Resolved Optical Remote Sensing of the Atmosphere, Springer Series in Optical Sciences, Springer, Heidelberg.*

Ansmann A., Ingmann, P., Le Rille, O., et al. (2006). Particle Backscatter and Extinction Profiling with a Spaceborne HSR Doppler Wind Lidar ALADIN, 2006, *in Reviewed and Revised Papers Presented at the 23rd International Laser Radar Conference (ILRC 2006), Nara, Japan, 24-28 July 2006, Editors Chikao Nagasawa, Nobuo Sugimoto, Part II,* 1015-1018.

Antonioly T. & Benetti, P. (1983). Study of afterpulse effects in photomultipliers, *Rev. Sci. Instrum.*, 54, 1777-1780.

Arshinov Y. F., Bobrovnikov, S.M., Zuev, V.E., et al. (1983). Atmospheric temperature measurements using a pure rotational Raman lidar, *Appl. Opt.* 22, 2984-2990.

Arshinov Yu. F. & Bobrovnikov, S.M. (1999). Use of a Fabry-Perot interferometer to isolate pure rotational Raman spectra of diatomic molecules *Appl. Opt.* 38, 4635-4638.

Arshinov Yu. F., Bobrovnikov, S.M., Serikov, I.B., et al. (1996). Calibration of a Raman-Lidar Gas Analyzer of Atmospheric Emissions from Plant Stacks Using a Remote Gas Chamber, *in Advances in Atmospheric Remote Sensing with Lidar, Selected papers of the 18th International Laser Radar Conference (ILRC) Berlin, 22-26 July 1996,* 427-430.

Arshinov Yu.F., Bobrovnikov, S.M., Nadeev, A.I., et al. (2002). Observation of range-resolved rovibrational Raman spectra of water in clear air and in a cloud with a 32-spectral-channel Raman lidar, *in Lidar Remote Sensing In Atmospheric and Earth Sciences - Reviewed and revised papers presented at the twenty-first International Laser Radar Conference (ILRC21), Québec, Canada, 8-12 July 2002, Luc R. Bissonnette, Gilles Roy, and Gille Vallée, Editors, Defence R&D Canada – Valcartier, Val-Bélair, Québec, Canada,* 31-34.

Arshinov Yu., Bobrovnikov, S., Serikov, I., et al. (2004). Optic-fiber scramblers and a Fourier transform lens as a means to tackle the problem on the overlap factor of lidar, *in Reviewed and Revised papers presented at the 22nd International Laser Radar Conference (ILRC 2004), 12-16 July, Matera, Italy, (ESA SP-561, June 2004),*.227-231.

Arshinov Yu., Bobrovnikov, S., Serikov, I., et al. (2005). Daytime operation of a pure rotational Raman lidar by use of a Fabry-Perot interferometer, *Appl. Opt..*, 44, 3593-3603.

Avila G.J., Fernández, J.M., Maté, B. et al. (1999). Ro-vibrational Raman Cross Sections of Water Vapor in the OH Stretching Region , *J. Mol. Spectrosc.*, 196, 77-92.

Balin I., Simeonov, V., Serikov, I., et al. (2004). Simultaneous measurement of temperature, water vapor, aerosol extinction and backscatter by Raman lidar, in Reviewed and Revised papers presented at the 22nd International Laser Radar Conference (ILRC 2004), 12-16 July, Matera, Italy, (*ESA SP-561*, June 2004); 139-143.

Balin I., Serikov, I., Bobrovnikov, S., et al. (2004a). Simultaneous measurement of temperature, water vapor, aerosol extinction and backscatter by Raman lidar, *Appl. Phys.*, B79, 775–782.

Balin Y.S., Samoilova, S.V., Krekova, M.M., et al. (1999). Retrieval of cloud optical parameters from space-based backscatter lidar data, *Appl. Opt.*, 38, 6365-6373.

Bauer H., Bauer, H.S. Wulfmeyer, V., et al. (2004). End-to-end simulation of the performance of WALES: Forward model, in Reviewed and Revised Papers Presented at the 22nd International Laser Radar Conference (ILRC 2004), Matera, Italy, 12-16 July 2004, Editors G. Pappalardo and A. Amodeo, *ESA-SP-561, Vol. II*, 1011-1014.

Beekmann M., Ancellet, G. Megié, G., et al. (1994). Intercomparison campaign of vertical ozone profile including electrochemical sondes of ECC and Brewer-Mast type and ground-based UV-differential absorption lidar, *J. Atmos. Chem.*, 19, 259-288.

Behrendt A., & Reichardt, J. (2000). Atmospheric temperature profiling in the presence of clouds with a pure rotational Raman lidar by use of an interference-filter-based polychromator, *Appl. Opt.*, 39, 1372-1378.

Behrendt A., Nakamura, T., Onishi, M, et al. (2002). Combined Raman Lidar for the Measurement of Atmospheric Temperature, Water Vapor, Particle Extinction Coefficient, and Particle Backscatter Coefficient, *Appl. Opt.*, 41, 7657-7666.

Behrendt A., Nakamura, T. &. Tsuda, T. (2004). Combined Temperature Lidar for Measurements in the Troposphere, Stratosphere, and Mesosphere, *Appl. Opt.*, 43, 2930-2939.

Behrendt A. (2005). Temperature Measurements with Lidar, p.p. 273-305, in Weitkamp C., (Editor), Lidar: Range-Resolved Optical Remote Sensing of the Atmosphere, Springer Series in Optical Sciences, Springer, Heidelberg.

Beyerle, G., Deckelmann, H., Neuber, R., et al. (2001). Occurrence of solid particles in the winter polar stratosphere above the nitric acid trihydrate co-existence temperature inferred from ground-based polarization lidar observations at Ny-Ålesund, Spitsbergen, J. Geophys. Res., 106, 2979-2992.

Bischel W. K. & Black, G. (1983). Wavelength dependence of Raman scattering cross-sections from 200-600nm, in Eximer Lasers – 1983 (OSA, Lake Tahoe, Nevada) editrs Rhodes C. K., H. Egger and H. Pummer, AIP Conference Proceedings, No. 100, Subseries on Optical Science and Engineering, No. 3. American Institute of Physics, New York. pp. 181-187.

Blum U., Fricke, K.H., Müller, K.P., et al. (2005). Long-term lidar observations of polar stratospheric clouds at Esrange in northern Sweden, Tellus B, 57, 412-422.

Bobrovnikov S.M., Arshinov, Yu.F., Serikov, I.B., et al. (2002). Daytime temperature profiling in the troposphere with a pure rotational Raman lidar, in: Lidar Remote Sensing In Atmospheric and Earth Sciences - Reviewed and revised papers presented at the twenty-first International Laser Radar Conference (ILRC21), Québec, Canada, 8-12 July 2002, editors Luc R. Bissonnette, Gilles Roy, and Gille Vallée, Defence R&D Canada – Valcartier, Val-Bélair, Québec, Canada, 2002, 717-720.

Böckmann C., Wandinger, U., Ansmann, A., et al. (2004). Aerosol Lidar Intercomparison in the Framework of the EARLINET Project. 2. Aerosol Backscatter Algorithms, *Appl. Opt.*, 43, 977-989.

Bohren C.F. & Huffman, D.R. (1983). Absorption and scattering of light by small particles", Wiley, New York.

Bösenberg J., Ancellet, G., Apituley, A., et al. (1993). Tropospheric ozone lidar intercomparison experiment, TROLIX'91, field phase report (report 102),. Max-Plank Instutut für Meteorologie, Hamburg, Germany. 24-27.

Bösenberg J. (1998). "Ground-based Differential-Absorption Lidar for Water Vapor and Temperature Profiling: methodology", Appl. Opt., 37, 3845-3860.

Bösenberg, J., Alpers, M., Ansmann, A., et al. (2002). EARLINET: Establishing the European Aerosol Research Lidar Network, in Lidar Remote Sensing in Atmospheric and Earth Sciences, Proceedings of the 21st International Laser Radar Conference, Quebec City, Canada, Defence R&D Canada - Valcartier, Val-Belair, Quebec, Canada. (Eds.) L. Bissonette, G. Roy, G. Vallee, 293-296.

Bösenberg, J. & Linné, H. (2002). Laser remote sensing of the planetary boundary layer, *Met. Zeitschrift*, 11, 233-240.

Bösenberg, J. & Matthias, V. (2003). EARLINET: European Aerosol Research Lidar Network to establish an aerosol climatology". *Final report for the period of February 2000 to February 2003, (Contract EVRI-CTI999-40003).*

Bösenberg J. (2005). Differential-Absorption Lidar for Water Vapor and Temperature Profiling, 213-239. In Weitkamp C., (Editor), 2005, Lidar, Range-Resolved Optical Remote Sensing of the Atmosphere, Springer Series in Optical Sciences, Springer, Heidelberg.

Bribes J.L., Gaufrès, R., Monan, M., et al. (1976)., Raman band contours for water vapor as a function of temperature, *Appl. Phys. Lett.*, 28, 336-337.

Bristow M.P., Bundy, D.H. & Wright, A.G. (1995). Signal linearity, gain stability, and gating in photomultipliers: application to differential absorption lidars, *Appl. Opt.*, 34, 4437-4452.

Browell E., Wilkerson, T.D., & McIlrath, T.J. (1979). Water vapor differential absorption lidar development and evaluation, *Appl. Opt.*, 18, 3474-3483.

Browell E.V. (1982). Lidar measurements of tropospheric gases, *Opt. Eng.*, 21, 128-132.

Browell E.V., Carter, A.F., Shipley, S.T., et al. (1983). NASA multipurpose airborne DIAL system and measurements of ozone and aerosol profiles, *Appl. Opt.*, 22, 522- 534.

Browell E.V., Ismail, S. & Shipley, S.T. (1985). Ultraviolet DIAL measurements of O3 profiles in regions of patially inhomogeneous aerosols, *Appl. Opt.*, 24, 2827-2836.

Browell E.V. (1989). Differential absorption lidar sensing of ozone, *Proc. IEEE*, 77, 419-432.

Browell E.V., Butler, C.F., Ismail, S. et al. (1990). Airborne Lidar Observations in the Wintertime Arctic Stratosphere: Polar Stratospheric Clouds, *Geophys. Res. Lett.*, 17, 385-388.

Browell E.V., Butler, C.F., Fenn, M.A., et al. (1993). Ozone and Aerosol Changes During the 1991-1992 Airborne Arctic Stratospheric Expedition, *Science*, 261, 1155-1158.

Browell E.V., Ismail, S. & Grant, W.B. (1998). Differential Absorption Lidar (DIAL) measurements from air and space, *Appl. Phys.*, B67, 399-410.

Browell, E.V., Fenn, M.A., Butler, C.F., et al. (2001). Large-scale air mass characteristics observed over the remote tropical Pacific Ocean during March - April 1999: Results from PEM-Tropics B field experiment, *J. Geophys. Res.*, 106(D23), 32481-32502, 10.1029/2001JD900001.

Browell E.V., Hair, J., Butler, C.F., et al. (2003). Ozone, aerosol, potential vorticity, and trace gas trends observed at high-latitudes over North America from February to May 2000, *J. Geophys. Res.*, 108, 8369, doi:10.1029/2001JD001390.

Bruneau D., Cazeneuve, H., Loth, C., et al. (1991). Double-pulse dual-wavelength Alexandrite laser for atmospheric water vapor measurements, *Appl. Opt.*, 30, 3930-3937.

Bruneau D., Quaglia, Ph., Flamant, C. et al. (2001). Airborne lidar LEANDRE II for water-vapor profiling in the troposphere. I. System description, *Appl. Opt.*, 40, 3450-3461.

Cairo F., Di Donfrancesco, G., Adriani, A., et al. (1999). Comparison of various linear depolarisation parameters measured by lidar, *Appl. Opt.*, *38*, 4425-4432.

Cairo F., Adriani, A., Miterbini, M., et al. (2004). Polar stratospheric clouds observed during the Airborne Polar Experiment–Geophysica Aircraft in Antarctica (APE-GAIA) campaign, *J. Geophys. Res.* 109 (D7), D07204 10.1029/2003JD003930.

Campbell J., Sassen, K, Welton, E.J., et al. (2006). Continuous cloud lidar monitoring at South Pole Station: Analysis of PSC formation and denitrification potential during 2003, *in Reviewed and Revised Papers Presented at the 23rd International Laser Radar Conference (ILRC 2006), Nara, Japan, 24-28 July 2006, Editors Chikao Nagasawa, Nobuo Sugimoto, Part I,* pp. 573-576.

Candy B.H. (1985). Photomultiplier characteristics and practice relevant to photon counting, *Rev. Sci. Instrum.*, 56, 183-193.

Carnuth, W. & Trickl, T. (2000). Transport studies with the IFU three-wavelength aerosol lidar during the VOTALP Mesolcina experiment, *Atmos. Environment*, 34, 1425-1434.

Carnuth W., Kempfer, u. & Trickl, t. (2002). Highlights of the tropospheric lidar studies at IFU within the TOR project, *Tellus B*, 54, 163-185.

Chaikovsky A., Ivanov, A., Balin, Yu., et al. (2006). CIS-LINET – Lidar Network for Monitoring Aerosol and Ozone in CIS Regions, in Reviewed and Revised Papers Presented at the 23rd International Laser Radar Conference (ILRC 2006), Nara, Japan, 24-28 July 2006, Editors Chikao Nagasawa, Nobuo Sugimoto, Part I, pp. 671-672.

Chanin M.L. & Hauchecorne, A. (1984). Lidar Studies of Temperature and Density Using Scattering, *MAP Handbook,* 13, 87-89.

Chazette P., Pelon, J., Moulin, C., et al. (2001). Lidar and satellite retrieval of dust aerosols over the Azores during SOFIA/ASTEX, *Atmos. Environment*, 35, 4297-4304.

Chen, W.N., Chiang, C.W. & Nee, J.B. (2002). Lidar ratio and depolarisation ratio for cirrus clouds, *Appl. Opt.*, 41, 6470-6476.

Coates P.B. (1973). The origins of afterpulses in photomultipliers, *Appl. Phys.*, 6, 1159-1166.

Coates P.B. (1973a). A theory of afterpulse effects in photomultipliers and the prepulse height distribution, *Appl. Phys.*, 6, 1862-1869.

Comstock J.M., Ackerman T. & Mace, G.G. (2002). Ground-based lidar and radar remote sensing of tropical cirrus clouds at Nauru Islands: Cloud statistics and radiative impacts, *J. Geophys. Res.*, 107(D), 4714, doi: 10.1029/2002JD002203.

Cohen A., Cooney, J.A. & Geller, K.N. (1976). Atmospheric temperature profiles from lidar measurements of rotational Raman and elastic scattering, *Appl. Opt.*, 15, 2896- 2901.

Cooney J. (1970). Remote measurement of atmospheric water vapor profiles using the Raman component of laser backscatter, *Journal of Appl. Meteorol.*, 9, 182-184.

Cooney J. (1972). Measurement of Atmospheric Temperature Profiles by Raman Backscatter, *Journal of Appl. Meteorol.*, 11, 108-112.

Cooney J. & Pina, M. (1976). Laser radar measurements of atmospheric temperature profiles by use of Raman rotational backscatter, *Appl. Opt.*, 15, 602-603.

Cooney J. (1986). Lidar method of measurement of atmospheric extinction and ozone profiles, *Appl. Opt.*, 25, 2035-2036.

Cooney J. (1987). Acquisition of atmospheric extinction profiles by lidar: further notes, *Appl. Opt.*, 26, 3485-3489.

Cuomo, V., Di Girolamo, P., Pappalardo, G., et al. (1998). Lidar in Space Technology Experiment correlative measurements by lidar in Potenza, southern Italy, *J. Geophys. Res.*, 103(D10), 11455-11464, 10.1029/98JD00789.

Curran R.J. (1989). Satellite-Borne Lidar Observations of the Earth: Requirements and Anticipated Capabilities, *Proceedings of the IEEE*, 77, 478-490.

Daumont D., Brion, J., Charbonnier, J., et al. (1992). Ozone UV Spectroscopy I: Absorption Cross-section at Room Temperature, *Journal of Atmospheric Chemistry*, 15, 145-155.

Del Guasta M., Morandi, M., Stefanutti, L. et al. (1998). Lidar Observation of Spherical Particles in a -65° Cold Cirrus Observed Above Sodankylä (Finland) during SESAME, *J. Aerosol Science*, 29, 357-374.

De Tomasi, F.A.M., Tafuro, M.R. & Perrone, M.R., (2006). Height and seasonal dependence of aerosol optical properties over southeast Italy, J. Geophys. Res., 111, D10203, 10.1029/2005JD006779.

Dessler A.E.; Palm, S.P., Hart, W.D., et al. (2006). Tropopause-level thin cirrus coverage revealed by ICESat/Geoscience Laser Altimeter System, J. Geophys. Res., 111, D08203 10.1029/2005JD006586.

Di Sarra, A., Cacciani, M., Fiocco, G., et al. (2002). Lidar observations of polar stratospheric clouds over northern Greenland in the period 1990–1997, *J. Geophys. Res.*, 107(D12), 4152, doi:10.1029/2001JD001074.

Donovan D.P., Whiteway, J.A. & Carswell, A.I. (1993). Correction for non-linear photon-counting effects in lidar systems, *Appl. Opt.*, 32, 6742-6753.

Douglass, A.R., Schoeberl, M.R. Kawa, S.R., et al. (2001). Browell, A composite view of ozone evolution in the 1995 - 1996 northern winter polar vortex developed from airborne lidar and satellite observations, *J. Geophys. Res.*, 106(D9), 9879-9896, 10.1029/2000JD900590.

Dörnbrack A., Birner, T., Fix, A. et al. (2002). Evidence for inertia waves forming polar stratospheric clouds over Scandinavia, *J. Geophys. Res.*, 107, 8287, doi: 10.1029/2001JD000452.

Doutriaux-Boucher M., Pelon, J., Trouillet, V. et al. (1998). Simulation of satellite lidar and radiometer retrievals of a general circulation model three-dimentional cloud data set, *J. Geophys. Res.*, 103, 26,025-26,039.

Dubock P., Endmann, M. & Ingmann, P. (2006). Progress with ADM-Aeolus, the Spaceborne Doppler Wind Lidar ADM, 2006, in Reviewed and Revised Papers Presented at the 23rd International Laser Radar Conference (ILRC 2006), Nara, Japan, 24-28 July 2006, Editors Chikao Nagasawa, Nobuo Sugimoto, Part II, pp. 1011-1014.

Dupont E., Pelon, J. & Flamant, C. (1994). Study of the moist convective Boundary layer structure by backscattering lidar, *Boundary-Layer Meteorology*, 69, 1-25.

Durieux E., Fiorani L., Calpini B., et al. (1998). Tropospheric ozone measurements over the Great Athens Area during the MEDCAPHOT-TRACE campaign with a new shot-per-hot DIAL instrument. Experimental system and results, *Atmos. Environment*, 32, 2141-2150.

Edlen B. (1966). The refractive Index of Air, *Metrologia*, 2, 71-80.

Edner H., Fredriksson, K., Sunesson, A. et al. (1987). Mobile remote sensing system for atmospheric monitoring, *Appl. Opt.*, 26, 4330-4338.

Edner H., Fredriksson, K., Sunesson, A. et al. (1987a). *Appl. Opt.*, 26, 3183 - 3185.

Edner H., Faris, G.W., Sunesson, A., et al. (1989). Atmospheric atomic mercury monitoring using differential absorption lidar techniques, *Appl. Opt.* 28, 921- 929.

Egeback A.L., Fredriksson, K.A. & Hertz, H.M. (1984). DIAL techniques for the control of sulfur dioxide emissions, *Appl. Opt.* 23, 722- 729.

Ehret G., Kiemle, C., Renger, W. et al. (1993). Airborne remote sensing of tropospheric water vapor with a near-infrared differential absorption lidar system, *Appl. Opt.*, 32, 4534-4551.

Ehret, G., Fix, A., Weiß, V., et al. (1998). Diode-Laser-Seeded Optical Parametric Oscillator for Airborne Water Vapor DIAL Application in the Upper Troposphere and Lower Stratosphere, *Appl. Phys.*, B67, 427-431.

Ehret G., Hoinka, K.P., Stein, J., et al. (1999). Low stratospheric water vapor measured by an airborne DIAL, *J. Geophys. Res.* 104, 31,351-31,360. (1999JD900959).

Ehret G., Amediek, A., Esselborn, M. et al. (2006). Upper Tropospheric Water Vapor and Particles Measured in the Tropics by Airborne H2O-DIAL during TROCCINOX and SCOUT-O3, *in Reviewed and Revised Papers Presented at the 23rd International Laser Radar Conference (ILRC 2006), Nara, Japan, 24-28 July 2006, Editors Chikao Nagasawa, Nobuo Sugimoto, Part II.* 707-710.

Eloranta E. (2005). High-Spectral Resolution Lidar, p.p. 144-163, *in Weitkamp C., (Editor), Lidar: Range-Resolved Optical Remote Sensing of the Atmosphere, Springer Series in Optical Sciences, Springer, Heidelberg.*

Eltermann L. (1968). UV, visible and IR attenuation for altitudes to 50 km, *Air Force Cambridge Research Laboratories.*

Ertel K., Jansen, F., Matthias, V., et al. (2001). Shipborne Water Vapor DIAL Measurements in the Tropical Western Pacific during the Nauru99 Campaign, in Dabas A., C. Loth and J. Pelon (Editors), 2001, "Advances in Laser Remote Sensing", Selected papers presented at the International Laser Radar Conference (ILRC), Vichy, France, 10-14 July 2000; Edition de l'Ecole polytechnique, Palaiseau-Paris, pp., 321-324.

Ertel K. (2004). Application and Development of Water Vapor DIAL System, *Dissertatioin zur Erlangung des Doktorgrades der Naturwissenschaften im Fachbereich Geowissenschaften der Universität Hamburg, Hamburg 2004. (in English).*

Ertel K., Linne, H. & Bössenberg, J. (2005). Injection-seeded pulsed Ti:sapphire laser with novel stabilisation scheme and capability of dual-wavelength operation, *Appl. Opt.*, 44, 5120-5126.

Faris G. W. &. Copeland, R.A. (1997). Ratio of oxygen and nitrogen Raman cross sections in the ultraviolet", *Appl. Opt.*, 36, 2684-2685.

Fenn R.V. (1966). Correlation between atmospheric backscattering and meteorological visual range, *Appl. Opt.*, 5, 615-616.

Fernald F.G., Hermann, B.M. & Raegan, J.A. (1972). Determination of Aerosol Height Distributions by Lidar, *J. Applied Meteorology,* 11, 482-489.

Fernald F.G. (1984). Analysis of atmospheric Lidar observations: some comments, *Appl. Opt.*, 23, 652-653.

Ferrare R.A., Melfi, S.H., Whiteman, D.N. et al. (1992). Raman lidar measurements of Pinatubo aerosols over south-eastern Kansas during November-December 1991, *Geophys. Res. Lett.,* 19, 1599-1602.

Ferrare R.A., Melfi, S.H., Whiteman, D.H., et al. (1995). A Comparison of Water Vapor Measurements Made by Raman Lidar and Radiosondes, *Journal of Atmos. and Oceanic Technology*, 12, 1177-1195.

Ferrare R.A., Melfi, S.H., Whiteman, D.H., et al. (1998). Raman lidar measurements of aerosol extinction and backscattering, 1, Methods and comparisons, *J. Geophys. Res.*, 103, 19,663-19,672 (98JD01646).

Ferrare R.A., Melfi, S.H., Whiteman, D.H., et al. (1998a). Raman lidar measurements of aerosol extinction and backscattering, 2, Deriviation of aerosol real refractive index, single-scattering albedo, and humidification factor using Raman lidar and aircraft size distribution measurements, *J. Geophys. Res.*, 103, 19,673-19,690 (98JD01647).

Fierli, F., Hauchecorne, A. & Knudsen, B. (2001). Analysis of polar stratospheric clouds using temperature and aerosols measured by Alomar R/M/R lidar , *J. Geophys. Res.*, 106 , D20 24,127 (2001JD900062).

Fiocco G. & Grams G. (1964). Observations of aerosol layer at 20km by optical radar, *J. Atmos. Science*, 21, 323-324.

Fiocco G., Calisse, P., Cacciani, M. et al. (1999). ABLE: Development of an Airborne Lidar. *J. of Atmos. and Oceanic Technology*, 16, 1337-1344.

Fiorani L. (1996). Une première mesure Lidar combinée d'ozone et de vent, à partir d'une instrumentation et d'une méthodologie coup par coup, *Thèse No 1585, Ecole Polytechnique Fédérale de Lausanne. (in French)*.

Fiorani L., Calpini B., Jaquet L., et al. (1997). Correction scheme for experimental biases in differential absorption lidar tropospheric ozone measurements based on the analysis of sot per shot data samples, *Appl. Opt.*, 36, 6857-6863.

Fiorani L., Calpini, B., Jaquet, L., et al. (1998). A combined determination of wind velocities and ozone concentrations for a first measurement of ozone fluxes with a dual instrument during the MEDCAPHOT-TRACE campaign, *Atmospheric Environment*, 12, 2151-2159.

Fix A, Wirth, M., Meister, A. et al. (2002). Tunable ultraviolet optical parametric oscillator for differential absorption lidar measurements of tropospheric ozone, *Appl. Phys.,* B75, 153 - 163.

Flamant C. & Pelon, J. (1996). Atmospheric boundary layer structure over the Méditéranée during a Tramontane event, *Q. J. R. Meteorol. Soc.*, 122, 1741-1778.

Flamant C., Pelon, J., Flamant, P., et al. (1997). Lidar determination of the Entrainment zone thickness at   the top of the unstable marine atmospheric boundary layer, Boundary-Layer Meteorology, 83, 247-284.

Flamant C., Pelon, J., Chazette, P., et al. (2000). Airborne lidar measurements of aerosol spatial distribution and optical properties over the Atlantic Ocean during a European pollution outbreak of ACE-2, *Telus B*, 52, 662 – 677.

Flamant C., Koch, S., Weckwerth, T., et al. (2004). The life cycle of a Bore event over the US Southern Great Plains during IHOP-2002, *in Reviewed and Revised papers presented at the 22nd International Laser Radar Conference* (ILRC 2004), 12-16 July, Matera, Italy, Part II, (ESA SP-561, June 2004), 635-638.

Franke K., Ansmann, A., Müller, D. et al. (2001). One-year observations of particle lidar ratio over the tropical Indian Ocean with Raman lidar, *Geophys. Res. Lett.*, 28, 4559-4562.

Fredriksson K., Galle, B., Nystrom, K., et al. (1981). "Mobile lidar system for environmental probing," Appl. Opt. 20, 4181- 4189.

Fredriksson K.A. & Hertz, H.M. (1984). "Evaluation of the DIAL technique for studies on NO2 using a mobile lidar system," *Appl. Opt.*, 23, 1403-1411.

Frioud M., Mitev, V., Matthey, R., et al. (2003). Elevated aerosol stratification above the Rhine Valley under strong anticyclonic conditions, *Atmos. Environment*, 37, 1785-1797.

Frioud M. (2003). Application of backscatter lidar to determine the aerosol distribution above complex terrain, *Thèse, Université de Neuchâtel, Neuchâtel*.

Frioud M., Mitev, V., Matthey, R., et al. (2004). Variation of the aerosol stratification over the Rhine Valley during Foehn development: a backscatter lidar study, *Met.. Zeitschrift*, 13, 175-181.

Gardner Ch., Miller, M.S. & Liu, C.H. (1989). Rayleigh Lidar Observation of Gravity Wave Activity in the Upper Stratosphere at Urbana, Illinoi", *Journal of Atmospheric Sciences*, 46, 1838-1854.

Gill R., Geller, K., Farina, J., et al. (1979). Measurement of Atmospheric Temperature Profiles using Raman Lidar, J. *Appl. Meteorol.*, 18, 225- 227.

Gobbi G.P. (1995). Lidar estimation of stratospheric aerosol properties: surface, volume and extinction-to-backscatter ratio, *Journal of Geophysical Research*, 100, 11,219-11,235.

Gobbi G.P. (1998). Polarisation lidar returns from aerosols and thin clouds: A framework for the analysis, *Appl. Opt.*, 37, 5505-5508.

Gobbi G.P., Barnaba, F., Giorgi, R., et al. (2000). Altitude-resolved properties of a Sahara dust event over the Mediterranean, *Atmos. Environment*, 34, 5119-5127.

Gobbi G.P., Barnaba, F., Van Dingenen, R., et al. (2003). Lidar and in situ observations of continental and Sahara aerosol: closure analysis of particles optical and physical properties, *Atmos. Chem. and Physics*, 3, 445-477.

Godin S., Mégie, G., David, C., et al. (1994). Airborne lidar observation of mountain-wave-induced polar stratospheric clouds during EASOE, *Geophys. Res. Lett.*, 21, 1335-1338.

Godish T. (1991). Air Quality, *Lewis Publishers, Michigan*.

Goers U.B. (1995). Laser remote sensing of sulfur dioxide and ozone with the mobile differential absorption lidar ARGOS, *Opt. Eng.*, 34, 3097-3112.

Goldsmith J.E.M., Blair, F.H., Bisson, S.E., et al. (1998). Turn-key Raman lidar for profiling atmospheric water vapor, clouds and aerosol, *Appl. Opt.*, 37, 4979-4990.

Grant W.B., Browell, E.V., Butler, C.F., et al. (1997). LITE Measurements of Biomass Burning Aerosols and Comparison with Correlative Airborne Lidar Measurements of Multiple Scattering in the Planetary Boundary Layer, 1997, in Advances in Atmospheric Remote Sensing with Lidar. Selected Papers of the 18th International Laser Radar Conference (ILRC), Berlin, 22-26 June 1996. Editors A. Ansmann, R. Neuber, P. Rairoux, U. Wandinger, Springer, Berlin, p. 153-156.

Hauchecorne A. & Chanin, M.L. (1980). Density and Temperature Profiles Obtained by Lidar between 35 and 70km, *Geophys. Res. Lett.*, 7, 566-568.

Hauchecorne A., Chanin, M.L., Malique, C. et al. (1998). Preliminary results of the ALISSA cloud lidar on board the MIR Space Station, in Proceedings of 19th International Laser Radar Conference, 6-10 July 1998 Annapolis, USA, editors U. N. Singh, S. Ismail, G. Schwemmer (NASA/CP-1998-207671/PT2), Part II, 931-934.

Heintzenberg, J., Tuch, T., Wehner, B., et al. (2003). Arctic haze over central Europe, *Tellus B*, 55, 796-807.

Hélière A., Bézy, J.L., Lefebvre, A., et al. (2006). The ESA EarthCARE Mission: Mission Concept and Lidar Instrument Pre-Development, in Reviewed and Revised Papers Presented at the 23rd International Laser Radar Conference (ILRC 2006), Nara, Japan, 24-28 July 2006, Editors Chikao Nagasawa, Nobuo Sugimoto, Part II, pp. 1041-1044.

Herzberg G. (1950). Molecular Spectra and Molecular Structure. I. Spectra of Diatomic Molecules, 2nd edition, *D. van Nostrand Company, Inc., Toronto*.

Hinkley E.D. (1976). Editor, Laser Monitoring of the Atmosphere, *Springer Verlag, Berlin*.

Hoff R.M., Harwood, M., Sheppard, A., et al. (1997). Use of airborne lidar to determine aerosol sources and movement in the Lower Fraser Valley (LFV), BC, *Atmospheric Environment*, 31, 2123-2134.

Hoff R.M. & Strawbridge, K.B. (1997). LITE Observations of Anthropogenically Produced Aerosols, in Advances in Atmospheric Remote Sensing with Lidar. Selected Papers of the 18th International Laser Radar Conference (ILRC), Berlin, 22-26 June 1996. Editors A. Ansmann, R. Neuber, P. Rairoux, U. Wandinger, Springer, Berlin, p. 145-148.

Hofmann D.J., Oltmans, S.J., Komhyr, W.D., et al. (1994). Ozone loss in the lower stratosphere over the United States in 1992-1993: Evidence for heterogeneous chemistry on the Pinatubo aerosol, *Geophysical Research Letters*, 21, 65-68.

Hooper W.P. & Eloranta, E. (1986). Lidar measurements of wind in the planetary boundary layer: the method, accuracy and results from joint measurements with radiosonde and kytoon, *J. Clim. Appl. Meteorol.* 25, 990-1001.

Hooper W.P. & Gerber, H.E. (1986). Down looking lidar inversion constrained by ocean reflection and forward scatter of laser light, *Appl. Opt.*, 25, 689-697.

Immler F. & Schrems, O. (2003). Vertical profiles, optical and microphysical properties of Sahara dust layers determined by a ship-borne lidar, *Atmos. Chem. and Phys.*, 3, 1353-1364.

Immler F. & Schrems, O. (2006). Lidar observations of extremely thin clouds at the tropical tropopause, *in Reviewed and Revised Papers Presented at the 23rd International Laser Radar Conference (ILRC 2006), Nara, Japan, 24-28 July 2006, Editors Chikao Nagasawa, Nobuo Sugimoto, Part I*, pp. 547-550.

Ismail S. & Browell, E.V. (1994). "Recent Lidar Technology Developments and Their Influence on Measurements of Tropospheric Water Vapor", *J. Armos. And Ocean Tech.*, 11, 76-84.

Ismail S. & Browell, E.V. (1989). "Airborne and Spaceborne Lidar Measurements of Water Vapor Profiles: A sensitivity Analysis", *Appl. Opt.*, 28, 3603-3615.

Ismail S., Koch, G.D., Barnes, B.W., et al. (2004)., Technology developments for tropospheric profiling of CO_2 and ground-based measurements, in Reviewed and Revised Papers Presented at the 22nd International Laser Radar Conference (ILRC 2004), Matera, Italy, 12-16 July 2004, Editors G. Pappalardo and A. Amodeo, ESA-SP-561, Vol. I, pp. 65-68.

Iwasaka Y. (1986). Measurement of depolarisation of stratospheric partiscles by lidar – a case study on the disturbed stratospheric aerosol layer by the volcanic eruption of Mt. El Chichon, *J. Geomag. Geoelectr.*, 38, 729-740.

Iwasaki S.Y., Tsushima, Y., Shirooka, R., et al. (2004). Subvisual cirrus cloud observations using a 1064nm lidar, a 95GHz cloud radar and radiosondes in the warm pool region, *Geophys. Research Letters*, 31, doi: 10.1029/2003GL019377.

Iwasaki S., Matsui, I., Shimitzu, A., et al. (2006). Observation of subvisual cirrus clouds with a lidar in Tarawa, Kiribati, in Reviewed and Revised Papers Presented at the 23rd International Laser Radar Conference (ILRC 2006), Nara, Japan, 24-28 July 2006, Editors Chikao Nagasawa, Nobuo Sugimoto, Part I, pp. 601-604.

Jäger H. (2004). Transport of aerosol from the tropics to Northern midlatitudes after major volcanic eruptions as observed by lidar at Garmish-Partenkirchen, in Reviewed and Revised Papers Presented at the 22nd International Laser Radar Conference (ILRC 2004),

Matera, Italy, 12-16 July 2004, Editors G. Pappalardo and A. Amodeo, ESA-SP-561, Vol. II, pp. 563-566.

Jenkins T.E. (1987). Optical sensing techniques and signal processing, *Prentice/Hall International,* Englewoods Cliffs, N.J.

Karyampudi V.M., Palm, S.P., Reagen, J.A., et al. (1999). Validation of the Saharan Dust Plume Conceptual Model Using Lidar, Meteosat, and ECMWF Data, *Bull. Am. Meteorol. Soc.,* 80, 1045–1075.

Keckhut P., Wild, J.D., Gelman, M., et al. (2001). Investigation on long-term temperature changes in the upper stratosphere using lidar data and NCEP analysis, *J. Geophysical Res.,* 106, 7937-7944 (2000JD900845).

Keckhut P., Randel, W., Claud, Ch., et al. (2006). Temperature lidar network and SSU/NOAA synergy for the Middle Atmosphere monitoring, in Reviewed and Revised Papers Presented at the 23rd International Laser Radar Conference (ILRC 2006), Nara, Japan, 24-28 July 2006, Editors Chikao Nagasawa, Nobuo Sugimoto, Part I, pp. 581-584.

Kempfer U., Carnuth, W., Lotz, R., et al. (1994). A wide range ultraviolet lidar system for tropospheric ozone measurements: development and application, *Rev. Sci. Instruments,* 65, 3145-3164.

Kent, G.S., Trepte, C.R., Skeens, K.M., et al. (1998). LITE and SAGE II measurements of aerosols in the southern hemisphere upper troposphere, *J. Geophys. Res.,* 103, 19111-19128, 10.1029/98JD00364.

Kiemle C., Kästner, M. & Ehret, G. (1995). The convective boundary layer structure from lidar and radiosonde measurements during the EFEDA'91 campaign, *J. Atmos. Ocean. Technol.,* 12, 771-782.

Kiemle C., Ehret, G., Giez, A., et al. (1997). Estimation of boundary-layer humidity fluxes and statistics from airborne differential absorption lidar (DIAL), *J. Geophys. Res.,* 102, 29189-29203.

Klett J.D. (1981). Stable Analytical Inversion Solution for Processing Lidar Returns, *Appl. Opt.* 20, 211-220.

Klett J.D. (1985). Lidar inversion with variable backscattering to extinction ratios, *Appl. Opt.,* 24, 1638-1643.

Kölsch H.J., Rairoux, P., Wolf, J.P., et al. (1989). Simultaneous NO/NO 2 DIAL Measurement Using BBO crystals, *Appl. Opt.,* 28, 2052- 2056.

Kreipl, S., Mücke, R., Jäger, H., et al. (2000) Spectacular cases and Long-Range Ozone and Aerosol Transport, in Advances of Laser remote Sensing - Selected papers presented at the 20th International Laser Radar Conference, Vichy, France, 10-14 July 2000, ed. Dabas A., C. Loth and J. Pelon, pp. 455-458.

Kunz G.J. & de Leeuw, G. (1993). Inversion of Lidar signals with the slope method, *App. Opt.,* 32, 3249-3256.

Lazzarotto B., Frioud, M., Larchevêque, G., et al. (2001). Ozone and water-vapour measurements by Raman lidar in the planetary boundary layer: error sources and field measurements, *Appl. Opt.,* 40, 2985-2997.

Liu Z., Voelger, P. & Sugimoto, N. (2000). Simulations of the Observation of Clouds and Aerosols with the Experimental Lidar in Space Equipment System, *Appl. Opt.,* 39, 3120-3137.

Long D. A. (2002). The Raman Effect, *Wiley, New York.*

Maeda M. , Sibata, T. & Akiyoshi, H. (1990). Optimum wavelengths in solar-blind UV ozone lidars, *Jap. J. Appl. Phys.*, 29, 2843-2846.

Malicet J., Daumont, D., Charbonnier, J., et al. (1995). Ozone UV spectroscopy. II. Absorption Cross-section and temperature dependence, *Journal of Atm. Chemistry*, 21, 263-273.

Manatt S. L. & Lane, A.L. (1993). A compilation of the absorption cross-section of SO2 from 106 to 403nm, *J. Quant. Spectrosc. Radiat. Transfer*, 50, 267-276.

Marsh A.K.P., Mitchell, N.J. & Thomas, L. (1991). Lidar studies of stratospheric gravity wave spectra, *Planetary and Space science*, 39, 1541-1548.

Martucci G., Matthey, R., Mitev, V., et al. (2006). Detection of ultra-thin tropical cirrus during TROCCINOX – a case study performed by two airborne lidars, in Reviewed and Revised Papers Presented at the 23rd International Laser Radar Conference (ILRC 2006), Nara, Japan, 24-28 July 2006, Editors Chikao Nagasawa, Nobuo Sugimoto, Part I. 585-588.

Martucci, G., Matthey, R., Mitev, V., et al. (2007). Comparison between backscatter lidar and radiosonde measurements of the diurnal and nocturnal stratification in the lower troposphere, *J. Atmos. and Oceanic Technol.*, 24, 1231-1244.

Matthey, R., Mitev, V., Lazzarotto, B., et al. (1997). Depolarisation/backscatter lidar for stratospheric studies, *SPIE, Vol. 3104, Lidar Atmospheric Monitoring*, pp. 2-11.

Matthey, R., Cacciani, M., Fiocco, G., et al. (2003). Observations of aerosol and clouds with the ABLE and MAL lidars during the mid-latitude and Arctic ENVISAT validation campaigns, Proceedings of the 16th ESA Symposium on European Rocket and Balloon Programmes and Related Research, St. Gallen, Switzerland, 2-5 June 2003, ESA-SP 530, ISBN 92-9092-840-9, ISSN 0379-6566. pp.579-584.

Matthey, R., Schilt, S., Werner, D., et al. (2006). Diode laser frequency stabilisation for water-vapour differential absorption sensing, *Appl. Phys.*, B85, 477-485.

Matthias V., Freudenthaler, V., Amodeo, A., et al. (2004). Aerosol Lidar Intercomparison in the Framework of the EARLINET Project. 1. Instruments, *Appl. Opt.* 43, 961-976.

Matthias V., Balis, D., Bösenberg, J., et al. (2004). Vertical aerosol distribution over Europe: Statistical analysis of Raman lidar data from 10 European Aerosol Research Lidar Network (EARLINET) stations, *J. Geophys. Res.*, 109, No. D18, D18201, 10.1029/2004JD004638.

Mattis I., Ansmann, A., Althausen, D., et al. (2002). Relative-Humidity Profiling in the Troposphere with a Raman Lidar, *Appl. Opt.*, 41, 6451-6462.

Mattis I., Ansmann, A., Müller, D., et al. (2004). Multiyear aerosol observations with dual-wavelength Raman lidar in the framework of EARLINET, *J. Geophys. Res.*, 109, No. D13, D13203, 10.1029/2004JD004600.

McCormick, P.M. & Swissler, T.J. (1983). Stratospheric aerosol mass and latitudinal distribution of the El Chichon eruption cloud for October 1982, *Geophys. Res. Lett.*, 9, 877.880.

McCormick M.P., Trepte, C. & Kent, G.S. (1983a)., Spatial changes in the stratospheric aerosol associated with the North Polar Vortex, *Geophys. Res. Lett.*, 10, 941-944.

McCormick M.P., Winker, D.M., Browell, E.V., et al. (1993). Scientific Investigations Planned for the Lidar In-Space Technology Experiment (LITE), *Bull. Am. Meteorol. Soc*, 74, 205–214.

McCormick M.P., Thomason, LW. & Trepte, C.R. (1995). Atmospheric effects of the Mt Pinatubo eruption, *Nature*, 373, 399 – 404. doi:10.1038/373399a0.

McCormick M.P. (2005). Airborne and Spaceborne Lidars, p.p. 355-397 in Weitkamp C., (Editor), Lidar: Range-Resolved Optical Remote Sensing of the Atmosphere, Springer Series in Optical Sciences, Springer, Heidelberg.

McElroy J.L. & Smith, T.B. (1986). Vertical Pollutant Distribution and Boundary Layer Structure Observed by Airborne Lidar near the Complex Southern California Coastline, *Atmospheric Environment*, 20, 1555-1566.

Mc Gee, T. J., Gross, M., Ferrare, R., et al. (1993). Raman DIAL measurements of stratospheric ozone in the presence of volcanic aerosols, *Geophys. Res. Lett.*, 20, 955-958.

Measures R.M. (1984). Laser remote sensing, fundamentals and applications, *John Wiley&Sons*.

Mégie, G., Allain, J.Y., Chanin, M.L., et al. (1977). Vertical profile of stratospheric ozone by lidar sounding from the ground, *Nature*, 270, 329-331.

Melfi, S.H., Lawrence, J.D., Jr. & McCormick, M.P. (1969). Observation of Raman Scattering by water vapor in the atmosphere, *Appl Phys Lett.*, 15, 295-297.

Melfi, S.H. (1972). Remote measurements of the atmosphere using Raman scattering, *Appl. Opt.*, 11, 1605-1618.

Melfi S.H., Spinhirne, J.D., Chou, S.H., et al. (1985). Lidar Observation of Vertically Organized Convection in the Plnetary Boundary Layer over the Ocean, *Journal of Climate and Applied Meteorology*, 24, 806-821.

Melfi, S. & Whiteman, D. (1985). Observation of lower-atmospheric moisture structure and its evolution using a Raman lidar, *Bull. Am. Meteorol. Soc.*, 66, 1288-1292.

Melfi, S.H., Evans, K.D., Li, J., et al. (1997). Observation of Raman scattering by cloud droplets in the atmosphere, *Appl. Opt.*, 36, 3551-3559.

Menut, L., Flamant, C., Pelon, J., et al. (1999). Urban boundary layer height determination from lidar measurements over the Paris area, *Appl. Opt.,* 38, 945-954.

Menzies, R.T., Tratt, D.M. & Hunt, W.H. (1998). Lidar In-Space Technology Experiment Measurements of Sea Surface Directional Reflectance and the Link to Surface Wind Speed, *Appl. Opt.*, 37, 5550-5559.

Middleton, W.E.K. & Spilhaus, A.F. (1953). Meteorological Instrument, *University of Toronto Press, Toronto.*

Mitev, V., Grigorov, I., Simeonov, V.B., et al. (1988). Raman lidar and probing of atmospheric parameters, *Soviet Journal Atmospheric Optics (Optica Atmosphery)*, 1, 122-124. /in Russian/.

Mitev, V.M., Grigirov, I.V., Simeonov, V.B., et al. (1990). Raman lidar measurement of atmospheric extinction coefficient profiles, *Bulg. J. of Physics*, 17, 67-74.

Mitev V.M., Grigorov, I.V. & Simeonov, V.B. (1992). Lidar measurements of atmospheric aerosol extinction profiles: a comparison between two techniques – Klett inversion and pure rotational Raman scattering, *Appl. Opt.*, 31, 6469-6474.

Mitev, V, Matthey, R. & Makarov, V. (2002). Miniature backscatter lidar for cloud and aerosol observation from high altitude aircraft, *Recent Res. Devel. Geophysics, ISBN:81-7736-076-0, Research Signpost.* 4, pp. 207-223.

Mitev, V., Matthey, R., Frioud, M., et al. (2005). Backscatter lidar observation of the aerosol stratification in the lower troposphere during winter Bise: a case study, *Meteorologische Zeitschrift*, 14, 663-669.

Müller D., Wandinger, U. & Ansmann, A. (1999). Microphysical Particle Parameters from Extinction and Backscatter Lidar Data by Inversion with Regularization: Theory, *Appl. Opt.*, 38, 2346-2357.

Müller, D., Ansmann, A., Wagner, F., et al. (2002). European pollution outbreaks during ACE 2: Microphysical particle properties and single-scattering albedo inferred from multiwavelength lidar observations, *J. Geophys. Res.*, 107, 4248, doi:10.1029/2001JD001110.

Müller D., Mattis, I., Ansmann, A., et al. (2004). Closure study on optical and microphysical properties of a mixed urban and Arctic haze air mass observed with Raman lidar and Sun photometer, *J. Geophys. Res.*, 109, D13206, 10.1029/2003JD004200.

Müller M., Neuber, R., Massoli, P., et al. (2004a). Differences in Arctic and Antarctic PSC occurrenceas observed by lidar in Ny-Ålesund (79 N, 12 E) and McMurdo (78 S, 167 E)", *Atmospheric Chemistry and Physics (discussions)*, 4, 6837-6866.

Murayama T., Okamoto, H. Kaneyase, H., et al. (1999). Application of lidar depolarisation measurement in the atmospheric boundary layer: Effects of dust and sea-salt particles, *J. Geophys. Res.*, 104, 31,781-31,792.

Murayama, T., Sugimoto, Nobuo; et al. (2001). Ground-based network observation of Asian dust events of April 1998 in east Asia, *J. Geophys. Res.*, 106, 18,345 (2000JD900554).

Nagai T., Nagasava, Ch., Abo, M., et al. (2006). Development of CO2 DIAL System Using 1.6μm Absorption Band, in Reviewed and Revised Papers Presented at the 23rd International Laser Radar Conference (ILRC 2006), Nara, Japan, 24-28 July 2006, Editors Chikao Nagasawa, Nobuo Sugimoto, Part I, pp. 541-544.

Nedeljkovic, D.; Hauchecorne, A. & Chanin, M.-L. (1993). Rotational Raman lidar to measure the atmospheric temperature from the ground to 30 km, *IEEE Trans. of Geophyiscs and Remote Sensing*, 31, 90-101.

Nevzorov A., Zuev, V. Burlakov, V., et al. (2004). Climatology of background stratospheric aerosol over Siberian regions according to lidar measurement data, in Reviewed and Revised Papers Presented at the 22nd International Laser Radar Conference (ILRC 2004), Matera, Italy, 12-16 July 2004, Editors G. Pappalardo and A. Amodeo, ESA-SP-561, Vol. II, pp. 601-604.

Newell, R.E., Zhu, Y., Browell, E.V., et al. (1996). Upper tropospheric water vapour and cirrus: Comparison of DC-8 observations, preliminary UARS microwave limb sounder measurements and meteorological analyses, *J. Geophys. Res.*, 101, 1931-1941.

Nyeki S., Kalberer, M., Colbeck, I., et al. (2000). Convective Boundary Layer Evolution to 4 km asl over High-Alpine Terrain: Airborne Lidar Observations in the Alps, *Geophys. Research Lett.*, 27, 689-692.

Omar A.H. & Gardner, C.S. (2001). Observations by the Lidar In-Space Technology Experiment (LITE) of high-altitude cirrus clouds over the equator in regions exhibiting extremely cold temperatures, *J. Geophys. Res.*, 106, 1227-1236.

Osborn, M.T., Kent, G.S. & Trepte, C.R. (1998). Stratospheric aerosol measurements by the Lidar in Space Technology Experiment, *J. Geophys. Res.*, 103, 11447-11454, 10.1029/97JD03429.

Overbeck, J.A.., Salisbury, M.S., Mark, M.B., et al. (1995). Required energy for a laser radar system incorporating a fiber amplifier or an avalanche photodiode, *Appl. Opt.*, 34, 7724-7730.

Palm S. P., Melfi, S.H. & Carter, D. (1994). New airborne scanning lidar system: application for atmospheric remote sensing, *Appl. Opt.*, 33, 5674-5681.

Palm S., Hlavka, D., Hart, W., et al. (2004). Calibration of the Geoscience Laser Altimeter System (GLAS) atmospheric channels, in Reviewed and Revised Papers Presented at the 22nd International Laser Radar Conference (ILRC 2004), Matera, Italy, 12-16 July 2004, Editors G. Pappalardo and A. Amodeo, ESA-SP-561, Vol. II, pp. 1003-1006.

Papayannis, A., Ancellet, G., Pelon, J., et al. (1990). Multiwavelength lidar for ozone measurements in the troposphere and lower stratosphere, *Appl. Opt.*, 29, 467-476.

Pappalardo G., Amodeo, A., Pandolfi, M., et al. (2004). Aerosol Lidar Intercomparison in the Framework of the EARLINET Project. 3. Raman Lidar Algorithm for Aerosol Extinction, Backscatter, and Lidar Ratio, *Appl. Opt.*, 43, 5370-5385.

Pelon, J., Flamant, C., Chazette, P., et al. (2002). Characterization of aerosol spatial distribution and optical properties over the Indian Ocean from airborne LIDAR and radiometry during INDOEX'99, *J. Geophys. Res.*, 107, 8029, doi:10.1029/2001JD000402.

Penndorf, R. (1957). Tables of refractive index for standard air and the Rayleigh scattering coefficient for the spectral region 0.2 to 20 μm and their application to atmospheric optics, J. of the Opt. Soc. of America, 47, 176-182.

Peneney, C.M., St. Peters, R.L. & Lapp, M. (1974). Absolute rotational Raman cross-sections for N_2, O_2 and CO_2, J. of the Opt. Soc. of America, 64, 712-716.

Penney C.M. & Lapp, M. (1976). Raman-scattering cross section for water vapor, J. of the Opt. Soc. of America, 66, 422-425.

Placzek E. (1934). Rayleigh-streung und Raman-effekt, in Handbuch der der radiology, editor G. Marx, p. 205-233.

Platt, C.M.R., Scott, J.C. & Dilley, A.C. (1987). Remote Sounding of High Clouds. Part VI: Optical Properties of Midlattitude and Tropical Cirrus, *Journal of the Atmospheric Sciences*, 44, 729-747.

Poberaj, G., Fix, A., Assion, A., et al. (2002). Airborne All-Solid-State DIAL for Water Vapour Measurements in the Tropopause Region, *Appl. Phys. B*, 75, 165-172.

Poole, L.R. & McCormick, M.P. (1988). Airborne lidar observations of Arctic polar stratospheric clouds: implications of two distinct growth stage, *Geophys. Res. Lett.*, 15, 21-23.

Proffitt M. H. & Langford, A.O. (1997). Ground-based differential absorption lidar system for day or night measurements of ozone throughout the free troposphere, *Appl. Opt.* 36, 2568-2585.

Reichardt J., Bisson, S.E., Reichardt, S., et al. (2000). Rotational vibrational-rotational Raman differential absorption lidar for atmospheric ozone measurements: methodology and experiment, *Appl. Opt.*, 39, 6072-6079.

Reitebuch, O., Chinal, E., Durant, Y., et al. (2004). Development of an airborne demonstrator for ADM-AEOLUS and campaign activities, in Reviewed and Revised Papers Presented at the 22nd International Laser Radar Conference (ILRC 2004), Matera, Italy, 12-16 July 2004, Editors G. Pappalardo and A. Amodeo, ESA-SP-561, Vol. II, pp. 1007-1010.

Renauld D., Pourny, J.C. & Capitini, R. (1980). Daytime Raman-lidar measurements of water vapor, Opt. Lett., 5, 233-235.

Renauld D. & Capitini, R. (1988). Boundary-layer water vapor probing with a solar-blind Raman lidar, J. Atmos. Ocean. Technol., 5, 585-601.

Renger, W., Kiemle, C., Schreiber, H.G., et al. (1997). Correlative Measurements in Support of LITE Using the Airborne Backscatter Lidar ALEX, in Advances in Atmospheric Remote Sensing with Lidar. Selected Papers of the 18th International Laser Radar Conference (ILRC), Berlin, 22-26 June 1996. editors A. Ansmann, R. Neuber, P. Rairoux, U. Wandinger, Springer, Berlin, p. 165-168.

Rizi V., Redaelli, G., Visconti, G., et al. (1999). Trajectory Studies of Polar Stratospheric Cloud Lidar Observations at Sodankylä (Finland) during SESAME: Comparison with Box Model Results of Particle Evolution, Journal of Atmospheric Chemistry, 32, 165-181.

Rocadenbosch, F., Comeron, A. & Albiol, L. (2000). Statistics of the slope-method estimator, Appl. Opt., 39, 6049-6057.

Rotach, M.W., Vogt, R., Bernhofer, C., et al. (2005). BUBBLE - an Urban Boundary Layer Project, Theoretical Appl Climatol, 81(3-4), 231 - 261, DOI: 10.1007/s00704-004-0117-9.

Rothman, L.S., et al (31 authors) (2005). "The HITRAN 2004 molecular spectroscopic database", Journal of Quantitative Spectroscopy and Radiative Transfer, 96, 139 - 204, DOI 10.1016/j.jqsrt.2004.10.008

Russell P.B., McCormick, M.P., Swissler, T.J., et al. (1981). Satellite and correlative measurements of the stratospheric aerosol. II. Comparison of measurements made by SAM II, dustsondes and an airborne lidar, J. Atmos. Sci., 38, 1295-1312.

Russell P.B., Morley, B.B.M., Livingston, J.M., et al. (1982). Orbiting lidar simulations. 1: Aerosol and cloud measurements by an independent-wavelength technique, Appl. Opt. 21, 1541-1553.

Russell P.B. & Morley, B.M. (1982). Orbiting lidar simulations. 2: Density, temperature, aerosol, and cloud measurements by a wavelength-combining technique, Appl. Opt., 21, 1554-1563.

Salby, M. L. (1996). Fundamentals of Atmospheric Physics, Academic Press, Inc., New York.

Sassen K. & Liou, K.N. (1979). Scattering of polarized light by water droplet, mixed phase and ice crystal clouds: I. Angular scattering patterns, J. Atmos. Sci., 36, 838-851.

Sassen, K.; DeMott, P.J.; Prospero, J.M.; et al. (2003). Saharan dust storms and indirect aerosol effects on clouds: CRYSTAL-FACE results, Geophys. Res. Lett., 30, 1633, 10.1029/2003GL017371.

Sassen K. (2005). Polarization in Lidars, p.p. 19-42, in Weitkamp C., (Editor), Lidar, Range-Resolved Optical Remote Sensing of the Atmosphere, Springer Series in Optical Sciences, Springer, Heidelberg.

Schneider, W., Moortgat, G. K., Tyndall, G.S., et al. (1987). Absorption cross-sections of NO2 in the UV and visible region (200 – 700 nm) at 298 K , Journal of Photochemistry and Photobiology, 40, 195-217.

Schoulepnikoff L., V. Mitev, V. Simeonov, B. Calpini and H. van den Bergh, 1997, Experimental investigation of high-power single-pass Raman shifters in the ultraviolet with Nd:YAG and KrF lasers, Appl. Opt., 36, 5026-5043.

Schoulepnikoff, L., van den Bergh, H., Calpini, B., et al. (1998). Tropospheric Air Pollution Monitoring, Lidar, in Encyclopedia of Environmental Analysis and Remediation, Robert A. Meyers, Editor, John Wiley & Sons, Inc, 1998, (ISBN 0-471-11708-0). pp. 4873-4909

Schutz, B.E., Zwally, H.J., Shuman, C.A., et al. (2005). Overview of the ICESat Mission, *Geophys. Res. Lett.*, 32, L21S01, doi:10.1029/2005GL024009.

Schwemmer, G.K., Dombrowski, M., Korb, C.L., et al. (1987). A lidar system for measuring atmospheric pressure and temperature profiles, *Rev. Sci. Instrum.*, 58, 2226-2237.

Schwemmer, G., Palm, S.P., Melfi, S.H., et al. (1997). Retrieval of Atmospheric Boundary Layer Parameters from LITE and LASAL, in Advances in Atmospheric Remote Sensing with Lidar. Selected Papers of the 18th International Laser Radar Conference (ILRC), Berlin, 22-26 June 1996. editors A. Ansmann, R. Neuber, P. Rairoux, U. Wandinger, Springer, Berlin, p. 161-164.

Serikov I.B., Arshinov, Y.F., Bobrovnikov, S.M., et al. (2002). Distortions of the temperature profile of the atmosphere acquired with a pure rotational Raman lidar due to sphericity of the Fabry-Perot interferometer plates: in Lidar Remote Sensing in Atmospheric and Earth Sciences - Reviewed and revised papers presented at the twenty-first International Laser Radar Conference (ILRC21), Québec, Canada, 8-12 July 2002, editors Luc R. Bissonnette, Gilles Roy, and Gille Vallée, Defence R&D Canada – Valcartier, Val-Bélair, Québec, Canada, (2002), pp. 721-724.

Sicard, M., Pérez, C., Rocadenbosch, F., et al. (2006). Mixed-Layer Depth Determination in the Barcelona Coastal Area From Regular Lidar Measurements: Methods, Results and Limitations, *Boundary Layer Meteorology*, 119, 135-157.

Sherlock, V., Hauchecorne, A. & Lenoble, J.J. (1999). Methodology for independent calibration of Raman backscatter water-vapor lidar systems, *Appl. Opt.*, 38, 5816-5837.

Sherlock V., A. Garnier, A. Hauchecorne, Ph. Keckhut, 1999a, Implementation and validation of a Raman lidar measurement of middle and upper tropospheric water vapour, *Appl. Opt.,* 38, 5838-5850.

Shipley S.T., Tracy, D.H., Eloranta, E.W., et al. (1983). High spectral resolution lidar to measure optical scattering properties of atmospheric aerosols. 1: Theory and instrumentation, *Appl. Opt.*, 22, 3716-3724.

Simeonov, V., Mitev, V., van den Bergh, H., et al. (1998). Raman frequency shifting in a CH4 :H2 :Ar mixture pumped by the fourth harmonic of a Nd :YAG laser, *Appl. Opt.*, 37, 7112-7115.

Sivakumar V., Bencherif, H., Rao, P.B., et al. (2006). Rayleigh lidar observation of double stratopause structure, in Reviewed and Revised Papers Presented at the 23rd International Laser Radar Conference (ILRC 2006), Nara, Japan, 24-28 July 2006, Editors Chikao Nagasawa, Nobuo Sugimoto, Part I, 565-568.

Smullin L.O. & Fiocco, G. (1962). Optical echoes from the moon, Nature, 194, 1267-1267.

Spinhirne J. D., M. Z. Hansen, and L. O. Caudill, 1982, Cloud top remote sensing by airborne lidar, Appl. Opt., 21, 1564- 1571.

Spinhirne J.D., M.Z. Hansen, L. & Simpson, J. (1983). The Structure and Phase of Cloud Tops as Observed by Polarization Lidar, J. Appl. Meteorol, 22, 1319–1331.

Spinhirne J.D., Chudamani, S., Cavanaugh, J.F., et al. (1997). Aerosol and cloud backscatter at 1.06, 1.54 and 0.53 μm by airborne hard-target calibrated Nd:YAG/methane Raman lidar, *Appl. Opt.*, 36, 3475-3494.

Staehr W., Lahmann, W. Weitkamp, C. (1985). Range-resolved differential absorption lidar: optimization of range and sensitivity, *Appl. Opt.*, 24, 1950-1956.

Stefanutti L., Castagnoli, F., Del Guasta, M., et al. (1992). A Four Wavelength Depolarization Backscattering Lidar for Polar Stratospheric Cloud Monitoring, *Appl. Phys.*, B55, 13-17.

Stefanutti, L., Castagnoli, F., DelGuasta, M., et al. (1992a). The Antarctic ozone lidar system, *Appl. Phys.*, B55, 3-12.

Stefanutti L., Sokolov, L., Balestri, S., et al. (1999). The M-55 Geophysica as a Platform for the Airborne Polar Experiment, *J. of Atmos. and Oceanic Technology*, 16, 1303–1312.

Stein B., Wedekind, C., Wille, H., et al. (1999). Optical classification, existence temperatures and coexistence of different PSC types", *Journal of Geophysical Research*, 104, 23'983-23'993.

Steyn D.G., Baldi, M. & Hoff, R.M. (1999). The detection of mixed layer depth and entrainment zone thickness from lidar backscatter profiles, *J. Atmos. Oceanic Technol.* 16, 953-959.

Stohl, A. & Seibert, P. (1998). Accuracy of trajectories as determined from the conservation of meteorological tracers, *Q. J. R. Meteorol. Soc.*, 125, 1465-1484.

Strawbridge, K. B. & Hoff, R.M. (1996). LITE validation experiment along California's coast: Preliminary results, *Geophys. Res. Lett.*, 23, 73-76, 10.1029/95GL03338.

Stull, R.B. (1988). An introduction to Boundary Layer Meteorology, *Kluwer Academic Press*.

Summa, D., Di Girolamo, P., Bauer, H., et al. (2004). End-to-end simulation of the performance of WALES: Retrieval module, in Reviewed and Revised Papers Presented at the 22nd International Laser Radar Conference (ILRC 2004), Matera, Italy, 12-16 July 2004, Editors G. Pappalardo and A. Amodeo, ESA-SP-561, Vol. II, pp. 1015-1018.

Theopold, F.A. & Bösenberg, J. (1993). Differential Absorption Lidar Measurements of Atmospheric Temperature Profiles: Theory and Experiment, *Journal of Atmospheric and Oceanic Technology*, 10, 165–179.

Thomas, A., Borrmann, S., Kiemle, C., et al. (2002). In situ measurements of background aerosol and subvisible cirrus in the tropical tropopause region, *J. Geophys. Res.*, 107, 4763, doi:10.1029/2001JD001385.

Thomas G.E. (1991). Mesospheric clouds and the physics of the mesopouse region, *Rev. of Geophysics*, 29, 553-557.

Thomas L., Jenkins, D.B., Wareing, D.P., et al. (1987). Lidar observation of stratospheric aerosol associated with the El Chichon eruption, *Ann. Geophysicae*, 5A, 47-56.

Thomas, L., Cartwright, J.C. & Wareing, D.P. (1990). Lidar observation of the horizontal orientation of ice crystals in cirrus clouds, *Tellus*, 42B, 211-216.

Thomas L., Marsh, A.K.P., Wareing, D.P., et al. (1994). Lidar Observations of ice crystals associated with noctilucent clouds at middle latitudes, *Geophys. Res. Lett.*, 21, 385-388.

Toon O. B., E. V. Browell, S. Kinne, J. Jordan, 1990, An Analysis of Lidar Observations of Polar Stratospheric Clouds, *Geophys. Res. Lett.*, 17, 393-396.

Turco, R.P., Whitten, R.C. &. Toon, O.B. (1982). Stratospheric aerosols: Observation and Theory, *Review of Geophysics and Space Physics*, 20, 233-279.

Turner D.D., Ferrare, R.A., Heilman, L.A., et al. (2002). Automated Retrievals of Water Vapor and Aerosol Profiles from an Operational Raman Lidar, *Journal of Atmospheric and Oceanic Technology*, 19, 37–50.

Uchino O, M. P. McCormick, T. J. Swissler, and L. R. McMaster, 1986, Error analysis of DIAL measurements of ozone by a Shuttle excimer lidar, *Appl. Opt.*, 25, 3946-3951.

Uthe E.E. (1991). Elastic scattering, fluorescent scattering, and differential absorption lidar observations of atmospheric tracers, *Optical Engineering*, 30, 66-71.

US Standard Atmosphere 1976, *Handbook of Chemistry and Physics, 64th Edition,* pp. F155-F157.

Vaughan, G., Wareing, D.P., Thomas, L., et al. (1988). Humidity measurements in the free troposphere using Raman backscatter, *Q. J. R. Meteorol Soc.,* 144, 1471-1484.

Vaughan, G., Wareing, D.P., Pepler, S.J., et al. (1993). Atmospheric temperature measurements made by rotational Raman scattering, *Appl. Opt.*, 32, 2758-2764.

Vaughan G., Wareing, D.P., Jones, S.B., et al. (1994). Lidar measurements of Mt. Pinatubo aerosol at Aberystwyth from August 1991 through March 1992, *Geophys. Res. Lett.*, 21, 1315-1318.

Vesselovskii, I.A., Cha, H.K., Kim, D.H., et al. (2000). Raman lidar for the study of liquid water and water vapor in the troposphere, *Appl. Physics B*, 71, 113-117.

Veselovskii I., Griaznov, V., Kolgotin, A., et al. (2002). Angle- and size-dependent characteristics of incoherent Raman and fluorescent scattering by microspheres. 2. Numerical Simulation, *Appl. Opt.*, 41, 5783-5791.

Von Gossart G., Hoffmann, P., von Zahn, U., et al. (1996). Mid-lattitude noctilucent cloud observation by lidar, *Geophys. Res. Lett*, 23, 2919-2922.

Von Cossart, G., Fiedler, J., von Zahn, U., et al. (1997). Noctilucent clouds: One- and two-color observations, *Geophys. Res. Lett.*, 24, 1635-1638.

Von Cossart, G. Fiedler, J. & von Zahn U. (1999). Size distribution of NLC particles as determined from three-colos observations of NLC by ground-based lidar, *Geophys. Res. Lett.*, 26, 1513-1516, 1999.

Wakimoto, R.M. & McElroy, J.L. (1986). Lidar Observation of Elevated Pollutioin Layers over Los Angeles, *Journal of Climate and Applied Meteorology*, 25, 1583-1599.

Wandinger, U., Ansmann, A., Reichardt, J., et al. (1995). Determination of stratospheric aerosol microphysical properties from independent extinction and backscattering measurements with a Raman lidar, *Appl. Opt.* 34, 8315-8329.

Wandinger, U. (1998). Multiple-Scattering Influence on Extinction- and Backscatter-Coefficient Measurements with Raman and High-Spectral-Resolution Lidars, *Appl. Opt.*, 37, 417-427.

Wandinger U., Mattis, I., Tesche, M., et al. (2004). Air mass modification over Europe: EARLINET aerosol observations from Wales to Belarus, *J. Geophys. Res.*, 109, D24205, 10.1029/2004JD005142 /21

Wandinger U. (2005) Raman Lidar, pp. 241-271, in Weitkamp C., (Editor), 2005, "Lidar, Range-Resolved Optical Remote Sensing of the Atmosphere", Springer Series in Optical Sciences, Springer, Heidelberg.

Wang Z., Zhou, J., Hu, H., et al. (1996). Evaluation of dual differential absorption lidar based on Raman-shifter Nd:YAG or KrF laser for tropospheric ozone measurements, *Appl. Phys.*, B62, 143-147.

Wang Z., Whiteman, D., Demoz, B., et al. (2004). A new way to measure cirrus cloud ice water content by using ice Raman scatter with Raman lidar, in Reviewed and Revised Papers Presented at the 22nd International Laser Radar Conference (ILRC 2004), Matera, Italy, 12-16 July 2004, Editors G. Pappalardo and A. Amodeo, ESA-SP-561, Vol. I, pp. 321-324.

Weitkamp C. (Editor) (2005). Lidar: Range-Resolved Optical Remote Sensing of the Atmosphere, *Springer Series in Optical Sciences, Springer, Heidelberg.*

Werner C. (2005). Dopler Wind Lidar, p.p. 326-354, in Weitkamp C., (Editor), Lidar: Range-Resolved Optical Remote Sensing of the Atmosphere, Springer Series in Optical Sciences, Springer, Heidelberg.

Whiteman, D.N., Melfi, S.H. & Ferrare, R.A. (1992). Raman lidar system for the measurement of water vapor and aerosols in the Earth's atmosphere, *Appl. Opt.* 31, 3068-3082 (1992).

Whiteman, D.N., Murphy, W.F., Walsh, N.W., et al. (1993). Temperature sensitivity of an atmospheric Raman lidar system based on an XeF excimer laser, *Opt. Lett.* 18, 247- 249.

Whiteman, D.N., Walrafen, G.E., Yang, W.E., et al. (1999). Measurement of isobestic point in the Raman spectrum of liquid water using a backscattering geometry, *Appl. Opt.*, 38, 2614-2615.

Whiteman, D.N. (2003). Examination of the Traditional Raman Lidar Technique. I. Evaluating the Temperature-Dependent Lidar Equations, *Appl. Opt.*, 42, 2571-2592.

Whiteman, D. N. (2003a). Examination of the Traditional Raman Lidar Technique. II. Evaluating the Ratios for Water Vapor and Aerosols, *Appl. Opt.*, 42, 2593-2608.

Whiteman, D.N.; Russo, F.; Demoz, B.; et al. (2006). Analysis of Raman lidar and radiosonde measurements from the AWEX-G field campaign and its relation to Aqua validation, *J. Geophys. Res.*, 111, D09S09 10.1029/2005JD006429.

Whiteman, D.N., Demoz, B., Di Girolamo, P., et al. (2006a). Raman Water Vapor Lidar Measurements During the International H2O Project. I. Instrumentation and Analysis Techniques, *J. Atmos. Oceanic Technol.*, 23, 157-169.

Whiteman, D.N., Demoz, B., Di Girolamo, P., et al. (2006b). Raman Water Vapor Lidar Measurements during the International H2O Project. II. Case Studies, J. Atmos. Oceanic Technol., 23, 170-183.

Whiteway J. & Carswell, A. (1994). Rayleigh Lidar Observation of Thermal Structure and Gravity Waves Activity in the High Arctic during a Stratosphere Warming, *Journal of Atmospheric Sciences*, 51, 3122-3136.

Whiteway, J.A., Duck, T., Carswell, A.I., et al. (2006). Lidar on the PHOENIX Mars Mission, in Reviewed and Revised Papers Presented at the 23rd International Laser Radar Conference (ILRC 2006), Nara, Japan, 24-28 July 2006, Editors Chikao Nagasawa, Nobuo Sugimoto, Part II, pp. 981-984.

Wiegner, M., Emeis, S., Freudenthaler, V., et al. (2006). Mixing layer height over Munich, Germany: Variability and comparisons of different methodologies, *J. Geophys. Res.*, 111, D13201, doi: 10.1029JD006593.

Winker, D. M. & Trepte, C.R. (1998). Laminar cirrus observed near the tropical tropopause by LITE, *Geophys. Res. Lett.*, 25, 3351-3354, 10.1029/98GL01292.

Winker, D.M., Hunt, W.H. & Hostetler, C.A. (2004). Status and Performance of the CALIOP Lidar, *Proc. SPIE,* vol 5575, 8-15.

Winker, D.M., Pelon, J. & McCormick, M.P. (2006). Initial results from CALIPSO, in Reviewed and Revised Papers Presented at the 23rd International Laser Radar Conference (ILRC 2006), Nara, Japan, 24-28 July 2006, Editors Chikao Nagasawa, Nobuo Sugimoto, Part II, pp. 991-994.

Wirth M., G. Ehret, P. Mörl, W. Renger, 1994, Two-dimensional stratospheric distribution during EASOE, *Geophys. Res., Lett.*, 21, 1287-1290.

Wirth, M. & Renger, W. (1996). Evidence of large scale ozone depletion within the arctic polar vortex 94/95 based on airborne LIDAR measurements, *Geophys. Res. Lett.*, 23, 813-816, 10.1029/96GL00772.

Wirth, M., Tsias, A., Dörnbrack, A., et al. (1999). Model guided Lagrangian observation and simulation of mountain polar stratospheric clouds, *J. Geophys. Res.*, 104, 23,971-23,981.

Wulfmayer, V., Bösenberg J., Lehmann, J., et al. (1995). Injection –seeded alexandrite ring-laser: performance and application in water-vapor differential absorption lidar, *Opt. Lett.*, 20, 638-640.

Wulfmeyer, V. (1998). Ground-based differential absorption lidar for water-vapor and temperature profiling: development and specifications of a high-performance laser transmitter, *Appl. Opt.*, 37, 3804-3824.

Wulfmeyer, V. & Bösenberg, J. (1999). Ground-based differential absorption lidar for water-vapor profiling: accesment of accuracy, resolution and meteorological applications, *Appl. Opt.*, 37, 3825-3844.

Wulfmeyer, V. (1999). Investigation of turbulent processes in the lower troposphere with water vapour lidar and radar-RASS, *J. Atmos. Sci.*, 56, 1055-1076.

Wulfmeyer, V. (1999a). Investigation of humidity skewness and variance profiles in the convective boundary layer and comparison of the latter with large eddy simulation results, *J. Atmos. Sci.*, 56, 1077-1087.

Wulfmeyer, V. & Walther, C. (2001). Future performance of ground-based and airborne water-vapor differential absorption lidar. I. Overview and theory, *Appl. Opt.*, 40, 5304-5320.

Wulfmeyer, V. Walther, C. (2001a). "Future performance of ground-based and airborne water-vapor differential absorption lidar. II. Simulation of the precision of a near-infrared, high-power system", *Appl. Opt.*, 40, 5321-5336.

Wulfmeyer, V. et al (20 authors) (2003). Lidar Research Network for Water Vapor and Wind, *Meteorologische Zeitschrift*, 12, 5-24.

Young, A.T. (1981). On the Rayleigh-scattering optical depth of the atmosphere, *Journal of Applied Meteorology*, 20, 328-330.

Ziao, Y., Hardesty, R.M. & Post, M.J. (1992). Multibeam transmitter for signal dynamic range reduction in incoherent lidar systems, *Appl. Opt.*, 31, 7623-7632.

Zwally, H.J., Schutz, B., Abdalati, W., et al. (2002). ICESat's laser measurements of polar ice, atmosphere, ocean, and land, *Journal of Geodynamics*, 34, 405-455.

2. LASER APPLICATIONS IN WATER MONITORING

In: Laser Applications in Environmental Monitoring
Editors: L. Fiorani and F. Colao

ISBN 978-1-60456-249-1
© 2008 Nova Science Publishers, Inc.

Chapter 2.1

LIDAR BATHYMETRY

Timothy Kearns

Environmental Systems Research Institute (ESRI),
380 New York Street, Redlands, California, USA
E-mail: tkearns@esri.com, tel.: +1 – 909 793 2853

Lindsay MacDonald

Environmental Systems Research Institute (ESRI),
380 New York Street, Redlands, California, USA
E-mail: lmacdonal@esri.com, tel.: +1 – 909 793 2853

ABSTRACT

Airborne Lidar bathymetry is an evolving technology with several applications for environmental monitoring. By employing two wavelengths of the electromagnetic spectrum, near infrared and green, sea floor depths in clear waters can be measured from an aircraft platform. Essentially, the depth is determined from a difference measurement between the water surface and sea bottom. Airborne Lidar bathymetry has evolved into a laser scanning technology that incorporates motion sensors, GPS, imaging cameras, and even topographic Lidar. An airborne sensor can acquire dense, accurate, elevation measurements quickly and efficiently. In fact, Airborne Lidar bathymetry sensors can collect millions of data points per hour of operation.

This chapter presents the historical development of airborne bathymetric Lidar technology and scientific related aspects. Data processing and products generated from Lidar bathymetric data are dependent on the end use and customer requirements. Common data processing techniques used to create digital products such as triangular irregular networks, digital elevation models and intensity images are also presented in this chapter; followed by two case studies discussing environmental monitoring using airborne Lidar bathymetry. A discussion surrounding selected applications of lidar bathymetry is also included: hydrography, geomorphic characteristics, rapid response environmental assessments, security and defence, environment monitoring, and river surveying. The high resolution and high quality of information generated from these sensors produce digital products that are growing the list of Lidar bathymetry applications. As this technology continues to mature, the range of applications will increase respectively.

INTRODUCTION

Airborne lidar bathymetry is a technology and a science that leverages laser light to measure the depth of relatively clear, shallow water. Surveys are generally executed using helicopters and fixed-wing aircraft that can collect the depth data points with very high density and at a rate that conventional sonar methods cannot meet. Most systems can collect elevation and depth simultaneously, which allows the aircraft to straddle the land/water interface, providing a seamless capture of the coastline. Although hydrographic offices have been relatively slow to adopt lidar bathymetry as a data source for navigational purposes, its reliability is improving with time. Coastal engineers, geomorphologists and militaries alike have been keen to harness the bathymetric modelling capabilities of this technology. When measured against conventional sonar, it is more cost effective, and can attain information in the nearshore zone that boats often cannot.

The basic principle behind lidar bathymetry is straightforward. Laser light, consisting of infrared and green wavelengths of the electromagnetic spectrum, is pulsed from an aircraft in a precise pattern to the surface of the water below. The infrared light is reflected back to the aircraft from the water surface, while the green light travels through the water column. Although a portion of the signal is attenuated by the water itself, energy from the green light reflects off the seafloor and is captured by the lidar sensor onboard the aircraft. [Danson, 2006] The water depth is obtained from the time difference between the infrared and green laser reflections using a simple calculation that incorporates the properties of the water column along with system and environmental factors. Bathymetric lidar data is typically very dense, with millions of data points collected per hour of operation. Point spacing can vary from sub-meter to several meters, providing the ability to generate high resolution digital elevation models or supplemental sounding data for hydrographic purposes.

There are two fundamental approaches to implementing a lidar bathymetry solution on an airborne platform. Permanent installations on large aircraft enable long sorties to remote locations, while compact, transportable systems have been developed that can be deployed upon 'aircraft of opportunity' on short notice [Guenther, 2006]. Of the systems in use at the time of printing, there are three primary commercial manufacturers: Optech (Canada), Tenix LADS Corporation (Australia) and Airborne Hydrography (Sweden) [Danson, 2006].

Applications of lidar bathymetry are varied, and as this new technology evolves, its scope of use will increase. Some of the projects where lidar bathymetry has been useful are: monitoring of shoaling (becoming more shallow) in navigation channels, harbour and coastal security, sediment transport studies, coral reef management, fisheries management, storm damage assessment, storm surge modelling, and submarine pipeline planning and construction [Guenther, 2006]. These are just a few of the many examples of lidar bathymetry, several of which will be discussed in more detail later.

In this chapter, we discuss a brief historical timeline of the development and evolution of bathymetric lidar and the underlying technology and science. A discussion on data processing and applications of this technology follow. To conclude this chapter, two case studies are presented. The first demonstrates the application of lidar bathymetry for environmental monitoring of coastal shoreline before and after a major climatological event. The second demonstrates the utility of lidar bathymetry in a riverine environment.

HISTORICAL PERSPECTIVE

Overview

Lidar bathymetry evolved in three distinct phases. First was the inception and experimentation with new and exciting laser technology spanning from the late 1960s into the mid 1970s. In the latter half of the 1970s through the late 1980s, significant research and development was pioneered by academia, the private sector, and national governments. Ultimately, these activities led to the commercialization of airborne lidar bathymetry that continues to grow today, with continual advances in the technology through research and development by an active community.

Inception and Early Development

It wasn't long after the invention of the laser in the early 1960's that military researchers turned their attention from land uses and applications of this innovative technology to the sea. Initially, submarine detection was the primary focus of research that utilized a laser underwater [Ott, 1965; Sorenson, 1966]. In 1969, the University of Syracuse Research Centre published a paper that confirmed that shallow water bathymetry was possible using laser technology [Guenther, 2006; Hickman and Hogg, 1969]. Most initial work on bathymetry mapping with lasers concentrated on research surrounding two and three dimensional sea floor profiling. Australia, Canada, Sweden, United States and the former Union of Soviet Socialist Republics (USSR) were leaders in this domain [LaRocque and West, 1999]. Initial efforts were limited to ship-based applications of underwater lidar for profiling, but the United States Air Force successfully tested water penetration of lidar from a tower over the Gulf of Mexico in the early 1970's [Levis et al, 1973].

Laser Scanning Research and Development

Although laser profiling of the sub-aqueous zone proved to be successful, it wasn't until the development of swath mapping that interest in the technology grew dramatically. Australia built a lidar profiling system, WRELADS-1 (Weapons Research Establishment Laser Airborne Depth Sounder), and tested it in 1976 and 1977 [Clegg, 1978]. Subsequently, a scanning system called WRELADS-2 was developed and tested in the late 1970's [LaRoque and West, 1999]. The Canadian government agencies, Canadian Hydrographic Service and Canada Centre for Remote Sensing, teamed with the private-sector firm Optech to build functional profiling prototypes in the mid 1970's. The Defence Research Establishment in Sweden had also explored profiling lidar bathymeters, but later worked with Canada to test the Mark-2 profiling system with a scanning mirror [LaRoque and West, 1999]. The United States, through agencies such as the National Aeronautical and Space Administration (NASA), National Oceanic and Atmospheric Administration (NOAA), and the United States (US) Navy, designed and built a scanning bathymeter called the Airborne Oceanographic Lidar (AOL) [Guenther and Goodman, 1978] which is still in use today. Experience from the

AOL program led to the American development of another scanning system known as HALS (Hydrographic Airborne Laser Sounder) [Houck, 1980].

Commercialization and Technology Evolution

The most recent significant shift in the growth of this technology was the commercialization of lidar bathymeters. Canadian-based Optech delivered the LARSEN-500 to the Canadian Hydrographic Service in 1985, which marked the world's first commercial operational airborne lidar bathymeter. Three years later, the FLASH-I, designed and manufactured by the Defence Research Establishment in Sweden with Optech as a sub-supplier was also built. In 1988, Optech built and delivered the ALARMS (Airborne Laser Radar Mine Sensor) to the United States government [LaRoque and West, 1999]. Australia awarded two domestic companies, BHP Engineering and Vision Systems, to continue the work done by the WRE and to build a commercially available Laser Airborne Depth Sounder (LADS) for the Royal Australian Navy [LaRoque and West, 1999].

Throughout the 1990s and into the early 21st century, the evolution of various systems continued. Optech built and delivered the SHOALS-200 in 1994 (US Army Corps of Engineers), a SHOALS-1000 in 2003 (Japanese Coast Guard), and a SHOALS-1000T and SHOALS-3000TH to the United States government in 2003. Saab Dynamics designed and manufactured two Hawk Eye systems in 1995 (Sweden). Fugro Pelagos of San Diego, USA became the first commercial surveying company to take delivery of a lidar bathymeter by Optech, the SHOALS-1000T, in 2005. In Australia, Tenix LADS Corporation consumed Vision Systems and continued to develop and provide services with the LADS Mk-II, delivered in 1998. In 2004, the United Kingdom Hydrographic Office, Blom ASA and Airborne Hydrography AB formed Admiralty Coast Surveys and began leveraging the Hawk Eye II, built by Airborne Hydrography AB of Sweden and delivered in 2005.

Other nations have continued operational research into lidar bathymeters such as China with Green Oceanographic Lidar and Russia with GOI, Chaika and Makrel-II [LaRoque and West, 1999]. In 2001, NASA built the Experimental Advanced Airborne Research Lidar (EAARL) and has been using it ever since [Guenther, 2006]. At the time of writing, eight systems worldwide are currently being used for operational lidar bathymetric purposes, and more systems are in development by Tenix LADS Corp. and Optech [Guenther, 2006].

THE SCIENCE OF LIDAR BATHYMETRY

Overview

Lidar bathymetry uses laser light to transmit a pulse of energy from an aircraft to the surface below (Figure 1). Most modern systems employ two wavelengths as output beams: infrared (IR) and green. The IR wavelength acts as the water surface reflector. It provides the system with the distance between the aircraft and the surface of the water. The green wavelength penetrates into the water and although portions of it are attenuated in the water column, reflected light energy from the sea bottom is reflected back to receivers in the aircraft. The basic principle between the depth calculation is from a difference measurement between the water surface and sea bottom. There are several variables that are taken into account for this measurement: speed of light in air and water, systematic factors, turbidity and other water characteristics. In most cases, this measurement is further adjusted to represent water depth relative to a particular datum, such as mean sea level, mean lower low water, lowest astronomical tide, lowest low water large tide or geoid.

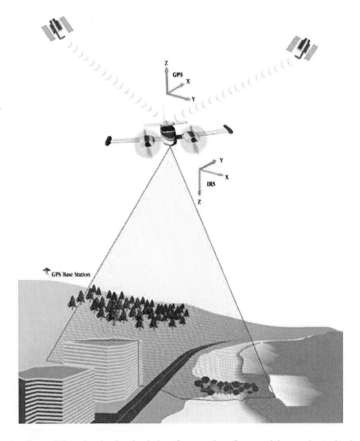

Figure 1. An illustration of the physical principle of operation for an airborne laser bathymeter.

Electromagnetic Energy

All forms of light energy are transmitted as electromagnetic energy. Electromagnetic energy encodes information as frequency, intensity, or polarization of the electromagnetic wave [Elachi, 1987]. Travelling at the speed of light, this energy, or radiation, can be reflected, absorbed, reradiated, or scattered from its source to the incident matter. The behaviour of the interaction of such energy with matter is governed by wavelengths. Wavelengths can vary from very short, such as gamma or cosmic rays to very long, such as radio waves. Gamma rays are measured on the order of Angstroms (10^{-10} m), while radio waves generally exceed 10 centimetres (cm). Most visible light wavelengths are measured in nanometres (nm).

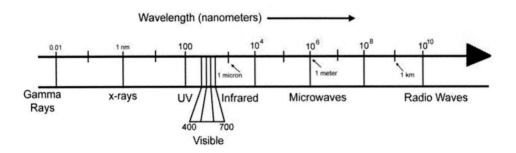

Figure 2. Electromagnetic spectrum

The electromagnetic spectrum is the continuum of energy that encompasses all wavelengths. It is divided into bands, or categories, of spectral regions (Figure 2). Starting with the longest of wavelengths and progressing to the shortest, the spectral bands are audio, radio, microwave, infrared, visible, ultraviolet, x-ray and gamma rays [Elachi, 1987]. Lidar bathymeters use green wavelengths (500-600 nm), from the visible band and infrared wavelengths (700-1100 nm), from the infrared band. It is this subtle difference in wave properties that governs how the electromagnetic energy interacts with the matter that it encounters when emitted from a lidar bathymetry system.

Although longer wavelengths tend to be absorbed by the water column, infrared light is strongly reflected at the surface and therefore, provides a good measure for detecting the sea surface returns [Guenther, 2006]. Shorter wavelengths, such as green light tend to attenuate less than other visible wavelengths in coastal waters [Guenther, 2006].

Waveform Characteristics

There are three primary components to the green wavelength return: Surface return, bottom return and volume backscatter (Figure 3). A variable portion of the transmitted green laser pulses are reflected at the air/water interface back to the optical receiver. In the proximity of the air/water interface, the return amplitude varies due to wave-slope statistics [Guenther, 2006]. As the green wavelength penetrates the water column, it interacts with the particulate matter and water molecules. This interference causes additional reflection back to the receiver and is termed volume backscatter [Guenther, 2006]. Below the air/water interface, volume backscatter is generally constant because turbidity of the water column does not tend to be localized, but rather, is evenly distributed over large areas of similar depth and characteristics. This reflection strength can vary from pulse to pulse and can possibly be mistaken as the surface return, resulting in an inaccurate depth measurement [Guenther and Mesick, 1988, Guenther, 1986]. Common applications of lidar bathymetry would generally not rely on a system that only transmits green electromagnetic energy due to the probability of inaccurate depth measurements [Guenther et al., 1994]. As depth increases, volume backscatter returning to the receivers decreases (Figure 3). When the green laser pulse encounters the sea bottom, energy is reflected back to the airborne receiver and the cycle of laser pulse(s), surface reflection, volume backscatter measurement, and sea bottom reflection is complete. This entire process, representing round-trip travel time in the water column, typically measures on the order of tens or hundreds of nanoseconds [Guenther, 2006]. Applying the speed of light in air and water and accounting for the beam nadir angle in water and the differences between surface returns and sea bottom returns, enables the calculation of the water depth at a specific geographic location. This measurement is normally further adjusted during post-processing to a specific chart, tidal or geodetic vertical datum.

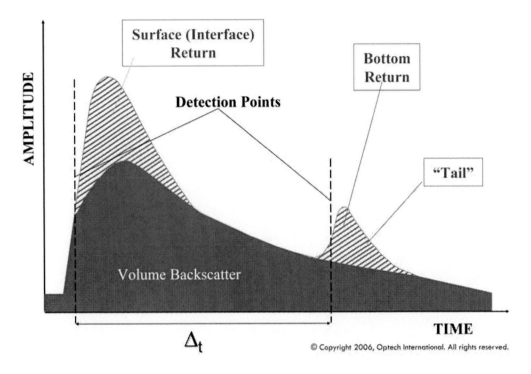

Figure 3. Volume backscatter of lidar waveform.

Laser Technology

At the heart of lidar bathymetry is the laser. Developed in the 1960s, laser technology is continually evolving and is a complex subject that is beyond the scope of this chapter. Only a brief description of laser technology relevant to lidar bathymetry is included.

Laser beams are commonly thought of as highly collimated, narrow beams of light. However, in lidar bathymetry, the laser pulses from the aircraft are intentionally broadened by several metres, in part due to eye safety considerations (Figure 4). There is still enough energy present to provide moderate signal to noise ratios and an eye-safe operational environment is achieved [Guenther, 2006]. Once the green waveform encounters the water column, refraction and beam spreading occur (Figure 4). Optical refraction is caused when the velocity of incident light that is transmitted through a surface is altered. In water, the speed of light is reduced, which decreases the wavelength but maintains the same frequency.

Traditional hydrographic surveying employs acoustic-electric transducers to transmit sound energy through the water column to the sea floor. Lidar bathymetry is similar in some respects to this competing technology. As with electro-acoustic waves emanating from traditional hydrographic echo sounding transducers, light energy in the water column is subject to spherical spreading and energy loss. With sonar systems the source of sound energy generates a sphere that grows from its origin. This spherical spreading causes the amplitude of the sound energy to decrease proportionally to the distance squared from the source.

Laser light energy in water behaves differently. Particulate matter and water molecules in the water column cause the light beam to spread and form a cone, a phenomenon called scattering. This cone's radius increases with the depth of penetration [Guenther, 2006]. For additional reading on this topic, the reader is directed to other sources information, such as "Water surface detection strategy for an airborne laser bathymeter" by Guenther and Thomas, 1990.

TECHNOLOGY OVERVIEW

Lidar bathymetry systems are considered active remote sensing systems and share similar characteristics to other remote sensing platforms. By definition, active remote sensing systems have their own energy source and transmit and receive a focussed portion of the electromagnetic spectrum. Both passive and active remote sensing systems share similar characteristics. They are comprised of a transmission source, energy medium(s), incident surface(s), energy-matter interactions (reflection, scattering, absorption, re-radiation) and reception of returned energy back to either a single or multiple receivers. Typically the incoming analogue signals are converted to a digital value and the data is stored on-board until further post-processing and download occurs. Most airborne lidar bathymetry systems consist of a laser, scanning mirror, amplifiers, telescope, optical filters, light detectors, digitizer, and digital storage system [Guenther, 2006].

Systems are mounted on either rotary-wing or fixed-wing aircraft. They can be used on aircraft of opportunity or can be permanently mounted. Collection altitudes vary from 200 to 500 m [Guenther, 2006]; point spacing and swath widths can be affected by altitude, if not accounted for by scan pattern. Aircraft velocity also affects the density of point spacing. Swath widths vary between systems but are typically 50-80% of the aircraft altitude. This is dependent on the scanning mirror capabilities, which currently range from 15° –22° [Guenther, 2006].

Two approaches are used for emitting laser light. The first uses two separate sources of radiation: one for green energy and the other for near infrared. The other approach is to emit infrared energy at 1064 nm and divert a portion of it to a crystal with non-linear optical properties. This doubles the frequency to 532 nm (green) and enables a simultaneous transmission of light from two wavelengths [Guenther, 2006]. NASA's EAARL differs from conventional lidar bathymeters in that it only utilizes a single wavelength of 532 nm to measure both air/water surface interface and sea bottom reflectance [Wright and Bock, 2002]. It does not use infrared light for the detection of the sea surface and uses a single laser that emits green light.

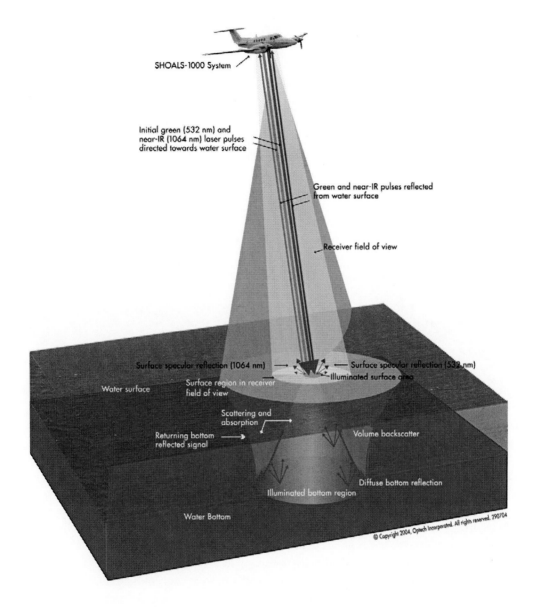

Figure 4. An illustration of the physical principle of operation for an airborne laser bathymeter (SHOALS-1000 depicted here).

There are three core enabling technologies for a complete lidar bathymetry system: a Global Positioning System (GPS), an Inertial Measurement Unit (IMU) and a laser transceiver. GPS measures the aircraft's position in the horizontal and vertical planes. An IMU measures aircraft attitude: roll, heave, pitch, and crab. A laser transceiver produces and transmits the laser, receives reflected infrared and green light energy and records the raw optical pulse measurements.

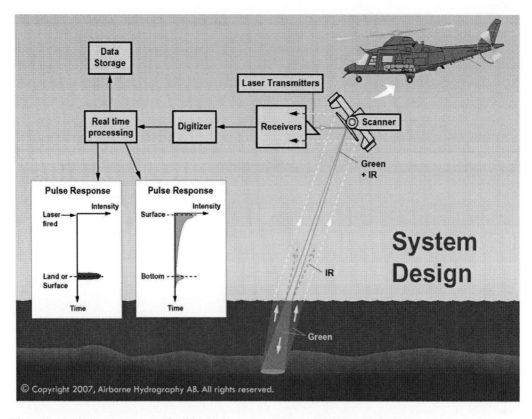

Figure 5. Typical components of a lidar bathymeter system (Hawk-Eye II depicted).

Global Positioning Systems

Knowing the accurate position of the aircraft is of paramount importance to the success of the entire system. Earlier methods, such as Loran-C, have been replaced with GPS, which employs satellite-based positioning with a high order of accuracy. GPS receivers use a constellation of earth-orbiting satellites to triangulate a position in x, y and z space. GPS technology has evolved significantly since lidar bathymetry's conception in the early 1970s.

There are three techniques predominantly employed by lidar bathymetry systems in operation today. Precise Code GPS (PGPS) is used infrequently and is slightly less accurate than Differential GPS (DGPS) [Guenther, 2006]. Kinematic GPS (KGPS) using carrier-phase post-processing techniques yields very accurate positions in both the horizontal and vertical planes. DGPS is the most commonly used and provides roughly 3 m or better accuracy in the horizontal plane. They are located on the aircraft, usually near the IMU and lidar bathymeter. The horizontal and vertical offsets between the GPS, IMU and lidar bathymeter must be precisely known so that all sub-systems are referenced to one common plane. These measurements can be obtained with a range-finding total station GPS or by other means [Guenther, 2006].

All three approaches require base station GPS receivers at locations on land with 'known' positions. Base stations measure local errors that translate into positional variations of the data. While DGPS and PGPS also require a vertical measurement of the water level at the time of survey, KGPS does not. KGPS can reference the WGS-84 ellipsoid which eliminates the need for local mean water ground control. By not referencing the local water level, thus removing a tidal variation to the data, KGPS is useful for surveying coastal areas where both topographic and bathymetric elevations are desired. In order to attain such high accuracy, especially vertically, multiple base stations are usually required. Since the maximum transmission range for base stations to a KGPS receiver is typically 10-30 km, survey costs can significantly increase over long coastline areas [Guenther, 2006].

PGPS is only in limited use by military organisations, whereas DGPS is widespread and is the most common among lidar bathymetry systems. The local errors measured by DGPS base stations are transmitted to users in-flight through high-, medium- or low-frequency means such as VHF (FM broadcast), beacons (fixed installations), and L-Band (geostationary satellites) [Guenther, 2006]. Subsequently, these corrections can be applied during post-processing procedures. For surveys that require local water level measurement (usually PGPS and DGPS), depths are reduced to a tidal datum that varies according to the end users' needs. Where confidence in predicted tidal data is low, increased accuracy may be achieved using a series of tidal level markers established prior to the survey for water level measurements at the time of acquisition.

Inertial Measurement Unit

The inertial measurement unit plays an important role in capturing the in-flight attitude variations of the aircraft. The IMU is usually made of a series of highly sensitive instruments such as gyros and accelerometers that measure the roll, pitch, and heading of the aircraft. The sampling rates are typically very high and can achieve a high precision. The IMU should be mounted close to the origin of the lidar pulse. Horizontal, vertical, and angular offsets must be determined within a very fine tolerance [Kletzli and Peterson, 1998], as any error in these offsets will propagate to the positional accuracy of the laser data.

Lidar Bathymeter

The lidar bathymeter is typically comprised of the sensor head, a control rack, and an operator display; however, this component breakdown varies across manufacturers. The laser itself is typically a diode pumped, pulsed, Nd:YAG laser [Wozencraft and Millar, 2005]. A scanning mirror scans across the swath width and in some systems; a constant angle is maintained through a tight integration with the IMU (Figure 6). Depending on the lidar bathymeter, there are either two laser beams generating and transmitting two wavelengths of light (IR and green), or a single laser beam generating at 1064 nm (IR) and a portion of energy derived from the main pulse frequency doubled to 532 nm (green). Regardless of the number of lasers, there is at least one receiver for each wavelength transmitted to the surface below. If the system is capable of topographic measurements there is usually a separate laser that measures land elevations. Most systems cannot collect topographic and bathymetric measurements simultaneously. NASA's EAARL and Airborne Hydrography AB's systems employ a double-laser approach to topography and bathymetry while Optech leverages the infrared beam for measuring the air/water interface as well as littoral relief.

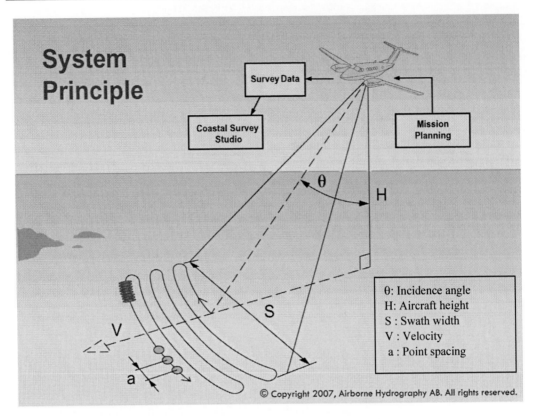

Figure 6. Principles of point spacing and scanning techniques from aircraft.

Horizontal Accuracy

Horizontal point spacing both at the sea surface and sea bottom varies due to a number of factors. Aircraft altitude, forward speed of aircraft, scanning design, water depth, and pulse frequency all contribute to how dense the data points will be. Older systems employed much lower frequencies, but most operational lidar bathymeters today operate on the order of 200-4000 Hz, which can result in upwards of 14 million soundings per hour. The point spacing ranges between 2 m and 6 m on the sea bed. Horizontal accuracy is determined by three primary factors: aircraft location, laser spot on water surface relative to the aircraft, and the sounding position on the sea bottom relative to the laser spot on the water surface [Guenther, 2006]. DGPS and KGPS using carrier-phase post-processing techniques has greatly aided this technology and now allows the precise location of the aircraft in x, y, z and T (time) space to be accurately measured. When coupled with an IMU, the attitude of the aircraft can be measured and any variations in the sensor position due to roll, pitch, and heading can be accounted for and applied to create an accurate position for the sounding. The beam position at the water surface is also an accurate measurement because it is a function of the nadir angle of the lidar beam and the aircraft altitude above the water surface [Guenther, 2006]. The sub-aqueous horizontal position of the sounding is much more difficult to accurately measure. It is subject to a host of other variables, including air/water interface characteristics (i.e. wind, waves, and wave slope), water column clarity, turbidity, and sea floor relief. While the uncertainties in aircraft or laser spot position result in horizontal positional inaccuracies, the effects of the water interacting with the light energy can affect each pulse separately. It is difficult to accurately estimate the error at the water surface, but some studies have shown that waves can affect the direction of the laser beam as it enters the water column. Wind speed, fetch, and the relative size of the laser spot diameter can contribute to increased horizontal error on a pulse-by-pulse basis [Guenther, 2006]. Scattering of light in the water column can also affect horizontal accuracy. This is largely a result of the turbidity of the water and the effects it has on scattering. Sea floor relief can affect both horizontal and vertical accuracy errors.

Vertical Accuracy

From a hydrographic perspective, the vertical accuracies attained by lidar bathymeters today are well within the constraints set forth by the International Hydrographic Organisation (IHO). All modern lidar bathymeters exceed the ±0.50 m (95% confidence level) vertical accuracy [Guenther, 2006]. Waters with high turbidity, steep sea floor relief, and small targets (i.e. mine-like objects or 1 metre cube blocks) affect the vertical accuracy of soundings. These factors can lead to local or systematic errors affecting large portions of the survey. The geometry of the sea floor affects both horizontal and vertical accuracy. Since each pulse has a finite beam width, on a steep sloping sea floor the tendency is to bias the sounding measurement towards the shoalest point in the beam. But in actuality, the true position is located horizontally down the slope [Guenther, 2006]. This leads to both a slight shift in the sampled vertical measurement as well as the horizontal one.

Integrating technologies

Sensor fusion is not limited to just GPS, IMU, and the laser unit. Newer systems have added high-resolution red/green/blue (RGB) or multispectral digital cameras that can collect digital imagery and also orthorectify it on the fly. This provides a tremendous capability when correlating lidar returns with both natural and anthropogenic features such as wharves, exposed rocks, aids to navigation, and other phenomena. Hyperspectral sensors have also been added, which provide water quality spectral characteristics, and in some cases, coral reef and sea bed signatures. Electromagnetic bandwidths typically range between 400 nm and 1050 nm, with a high spatial resolution from 25 cm to 1.5 m. Up to 288 bands of spectral information can be collected simultaneously with the lidar returns (Figure 7).

Data Processing

The goals of a lidar bathymetric survey vary depending on the application of the data collection. Common products from lidar bathymetry are: two-dimensional profiles, high resolution digital elevation models, contour isolines, seafloor characterisation from backscatter, and the post-processed soundings as discrete point data. The generic workflow that occurs for typical bathymetric lidar data is: (1) depth measurements collected and a partial solution derived in-flight, (2) auto post-processing to apply GPS/IMU refinement and waveforms analysis, (3) data validation through quality assurance/quality control process, (5) data transformation to derived products and, (6) seafloor reflectance data.

Figure 7. A simultaneous lidar bathymeter with digital camera and ITRES-CASI (hyperspectral imager) coverage (SHOALS system depicted here).

Data Collection

The lidar receiver measures not a single measurement, or even two single measurements. Rather, it measures and collects waveforms. It can vary from system to system, but there are typically two waveforms collected: one for the infrared signal and one for the green laser pulse. The bathymetric waveform is more complicated than topographic lidar because of the reflection of light from the water surface, particles in the water column and sea floor reflections [A. Axelsson, per. comm., 2007]. Complex algorithms, in combination with proprietary software, separate surface and sea bottom returns from each waveform, and the depths are calculated through waveform analysis. It is from these waveforms that peaks can be identified to indicate both surface and seafloor, but also the bottom reflectance can be derived [Wozencraft et al., 2007]. The lidar waveform is stored as raw data for further post-processing and analysis post-flight. However, a marginal amount of automated processing is done in-flight to provide the operator with a quality check of the waveform through a graphical display. [A. Axelsson, per. comm., 2007].

Auto Post-Processing

Although a limited amount of GPS and IMU information is used at the time of collection, a great deal of auto-processing is accomplished after the survey. This automated refinement of the sounding positions (both vertically and horizontally) varies from vendor to vendor, but each has similar characteristics. GPS (either DGPS or KGPS) measurements from reference stations are incorporated to further enhance the positioning precision of the sensor. This can increase the accuracy to just a few centimetres [A. Axelsson, per. comm., 2007]. Any adjustments to the depth solution from timing latency in scanner angles, attitude and altitude data are resolved and applied [Guenther et al., 2000]. Most of these activities occur within software applications provided by the lidar bathymetry vendor; the order with which these systems undertake certain corrections varies.

Filtering out errant points that are artifacts is usually part of the automated post-processing. Predicted tidal or measured water level information, is also merged with the GPS data to refine the sounding positions [C. Lockhart, per. comm., 2007]. Depths are referenced to the ellipsoid if KGPS is used as the primary vertical reference [Guenther et al., 2000]. One of the key roles of automated post-processing software is the discrimination of surface and bottom returns from the waveforms. There are various algorithms that exist to reject noise, artifacts, and false targets in the returned volume backscatter [Guenther et al., 2000]. Once the bottom and surface peaks are identified, the determination of the mean water level occurs. Subsequently, the removal of wave heights is automatically performed [Thomas and Guenther, 1990]. If KGPS is used for positioning, mean water surface and wave height removal is not required. The speed of light in water is not as crucial to depth determination as the speed of sound is with sonar depth measurements, but it still varies based on temperature, salinity, and water depth. Depending on the accuracy needs of the survey and the available field data available, speed of light is applied to the overall solution [Guenther et al., 2000].

It is important to note that bathymetric and topographic auto post-processing are significantly different. Lidar bathymetry systems must employ a more complex inverse transformation to account for the variability of how light behaves when traveling through the water column [A. Axelsson, per. comm., 2007].

Quality Control

The validation of lidar bathymeter depths takes place both in-flight and post-flight. A single operator monitors incoming data such as waveforms, error messages, system warnings, and pre-processed depths. Other survey characteristics such as track lines, survey area extents, and system-defined parameters are also presented to the operator through system specific software interfaces [Guenther et al., 2000]. Some aspects of data validation also include consistency checks to ensure waveforms are 'normal'. Data validation also includes the monitoring of warnings from abnormal readings that the in-flight operator detects. In addition to waveform checks, confidence levels are defined through a variety of potential error sources which enable a quantitative estimate of the depth accuracy for each pulse [Guenther et al., 2000].

Calibration of the sensor head itself is usually accomplished by a vigorous series of ground-checks, such as 'hard-target' range finding of a known distance and precise angle measurements of the slant-range and scanner azimuth angles [Guenther et al., 2000]. Post-flight quality control is partially managed by auto post-processing discussed previously. Subsequent to auto processing, most data providers employ manual techniques for final checks. Flights lines are typically flown with overlap, and cross-check lines provide a further means for post-flight operators to investigate discrepancies in the collected depth measurements. Shallow depths (0-1 m) tend to be problematic and are usually investigated manually. Various software tools exist, both proprietary and commercially available, to inspect sounding depths statistically and visually. Investigators can manually edit the data to remove artifacts and unwanted objects, or to verify wrecks, rocks and shoal areas [J. Wozencraft, per. comm., 2007].

Bathymetric lidar data contains a wide assortment of attributes associated with each point. These can include quality of position attributes, which can be extracted and used in quality assurance procedures [A. Axelsson, per. comm., 2007]. Once the vendor has met the objectives of the survey's requirements for quality control, the data is prepared for archive and distribution.

Data Deliverables

Depending on the end-use of the data, it is converted into either a propriety or open-source format for distribution. ASCII (American Standard Code for Information Interchange) text files are a common means of distributing tiled, or segmented, lidar data. A lidar exchange file (LAS) with an open-source format developed in part by the American Society of Photogrammetry and Remote Sensing Society may also be utilized for data distribution. Binary files, such as LAS, and proprietary formats by vendors are better suited to the large data volumes that are produced.

Data Transformation

The initial and post-processing procedures of lidar bathymetric data yield a cleaned, processed xyz sounding set of depths. There are few instances where the full, cleaned, sounding set of data is used at full density. Typically, it is decimated to a reduced data set that removes redundant data and makes the dataset more manageable for visualization, data management, and product generation. Commercially available or proprietary software applications are available; each with varying techniques to support data transformation. When the end result is support of hydrographic products such as Electronic Nautical Charts (ENCs), 'binning' is often applied. This process reduces the data to a regularly spaced grid, but 'golden' and shoal (shallow) soundings are retained [Guenther, 2006]. Since this technique preserves the shoalest depths, the data is tailored for nautical products where safety of navigation is of utmost importance. If the end resultant products are according to a fixed scale, such as printed paper charts, the full density data can be reduced by the most appropriate decimation factor using commercial map production software. Only in limited cases, such as engineering applications, or very large scale products, would the full density data set be used [Guenther, 2006].

In most applications of bathymetric lidar, the full density sounding set undergoes a data transformation. When transforming xyz point information into new features, they will adhere to a point-to-point, point-to-line, or point-to-area transformation (Table 1) [Bonham-Carter, 1999]. Subsequent to this initial transformation, they can be further modified to new points, lines, or areas (Figure 8). Points-to-areas transformations are very common and typically result in a Triangulated Irregular Network (TIN) (Figure 8). A TIN is a multi-faceted surface where each sounding in the input xyz data is included as a mass point in the creation of the TIN. Using either Delauney or Linear interpolation techniques, irregularly sized facets are created between each group of three points, which allows the calculation of a depth anywhere on the entire surface. Relative accuracy is maintained because no points are excluded from the resultant TIN. Dense data yields more accurate results, while sparse data can be problematic and may not accurately represent the sea floor surface.

Table 1. Geospatial data transformation

From/To	Points	Lines	Areas
Points	Interpolation	Contouring	Thiessen polygons, buffers, TINs Inverse Distance Weighting, Density Mapping
Lines	Line intersections	Line smoothing	Buffer zones
Areas	Sampling at points	Medial axis (called skeletonizing)	Overlay

Contour isolines are usually created from surfaces, rather than points themselves, and are a further by-product of bathymetric lidar data (Figure 8). Popular in use with hydrographic organizations, engineering applications, and the oil and gas industry, contour isolines aid with decision making for a variety of uses and are often a good way of communicating sea floor surface generalizations.

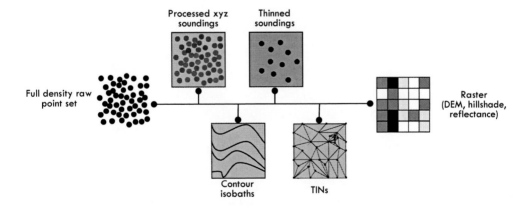

Figure 8. Data transformation highlighting examples of lidar bathymetry products.

Probably the most common method of viewing, analysing and modelling elevation data is with raster formats, specifically a Digital Elevation Model (DEM) (Figure 8). Also referred to as a Digital Terrain Model (DTM), these images are a regularly-spaced matrix of cells in rows and columns. Each cell, or pixel, in the grid holds a single value, representing either a positive or negative elevation. The representation size in horizontal units is usually dependent on the point spacing of the raw xyz sounding density, but is typically 2 to 6 m. DEMs can be created from points, lines, or areas and vary in their spatial resolution, accuracy, and interpolation technique. Creating an elevation model from points uses many algorithms; averaging a radius of points within a cell is one such example, 'binning' and taking the shoalest point in the cell is another. DEMs are often created from vector surfaces, such as TINs, Voronoi, or Thiessen areas [Bonham-Carter, 1999].

From an information content perspective, the DEM is a very valuable data source. However, it is often the derived products from DEMs that people readily associate with lidar bathymetric data. Slope and aspect models, shaded relief models, and three dimensional models provide a visual perspective that is difficult to attain by viewing the DEM in its native state (Figure 9).

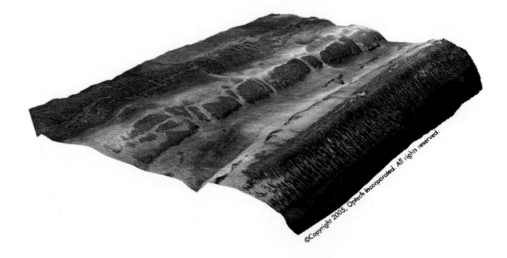

Figure 9. By integrating a DEM with sea floor reflectance, a clear picture emerges that depicts bathymetry. These depths near Port Everglades, USA range from 1m - 30m.

Seafloor Reflectance

Similar to topographic lidar, bathymetric lidar can also measure the intensity, or amount, of returned energy to the receiver. A portion of the electromagnetic energy that penetrates into the water column is lost due to attenuation, scattering, and absorption by the water column. The strong peaks in the return energy to the receiver indicate the interaction of green wavelengths with the sea floor and vary with differing absorption rates of the incident matter. Stronger energy returns suggest less energy absorption, while weaker energy returns suggest more absorption by the sea bed. These data help characterize the sea floor composite material and can be classified into different categories of superficial characteristics. Data is captured as a reflectance value and usually is digitally converted from the waveform to a normalized value between 0 and 255. Bathymetric reflectance data is typically represented in a grey-scale (8-bit) raster format. While it is difficult to fully classify sea floor reflectance data into bottom types, classification can be improved by integration with other sensor data, such as hyperspectral or RGB image data (Figure 10) [Tuell et al., 2005].

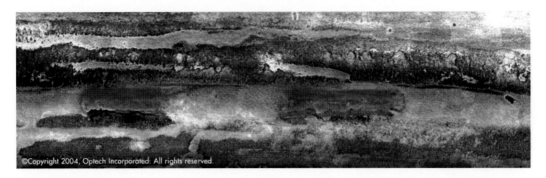

Figure 10. This is a SHOALS reflectance image of a 3 km x 5 km area of the sea-floor immediately south of the entrance channel to Port Everglades, USA. Clearly visible is the bottom morphology consisting of reefs paralleling the coast and separated by regions of sand, sea grass and mixed vegetation.

Applications

As an emerging technology that is still relatively immature, the full capabilities of lidar bathymetry are still unknown. Many of the initial efforts were designed for the hydrographic community and mapping of the nearshore zone in support of nautical charts. Lidar bathymetry has expanded to include the classification of sub-aqueous structures and geomorphology, rapid response environmental assessments, security and military applications, environmental monitoring, and river surveys.

Hydrography

The principal goal behind hydrographic surveying is to measure sea floor depths and hazards to navigation that are reflected on nautical charts and products. Because the safety of navigation to mariners is of utmost importance, both vertical and horizontal accuracies must adhere to high standards. The IHO is the accepted body that sets forth such standards, which must be adhered by hydrographic offices that produce nautical charts and ENCs. "Order-1" specifications from the IHO mandate that all sources, including tides, must meet ±0.50 m (95% confidence level) vertical accuracy [Guenther, 2006]. Most hydrographic offices strive to meet this accuracy for shallow water surveys. It is not until deeper soundings are attained that they may reduce accuracy thresholds to "Order-2" [Engstrom and Axelsson, 2001]. It is well known that lidar bathymetry systems meet IHO Order-1 accuracy specifications; all lidar bathymeter manufacturers will substantiate these claims. Many hydrographic agencies, such as the Royal Australian Navy, United Kingdom Hydrographic Office, Canadian Hydrographic Service, and the Swedish Maritime Administration have used lidar bathymeter sounding data on navigational products [Guenther, 2006]. One of the principal limitations of lidar bathymetry for hydrography is the detection of small objects. Due to the strong bottom return, it is difficult for the receiver to resolve the presence of small objects less than 1 cubic metre on the sea floor [Guenther, 2006]. This can be problematic as it is imperative that potential hazards to navigation, such as rocks or submerged vessels, are detected and made aware to the mariner. This challenge is shared by traditional multibeam echo-sounding survey techniques as well. In order to overcome this challenge, seafaring survey vessels typically use a minimum of 200% coverage at high spatial resolutions (2-4 m). Lidar surveys can also be modified to alleviate this concern through increased survey density and adjusted survey plans and parameters [Guenther, 2006].

Geomorphic Characteristics

Marine geologists, scientists and engineers are all principally interested in the geomorphological and superficial characteristics of the sea floor. Lidar bathymetry has enabled investigators to study sediment transport [Irish and Lillycrop, 1997; Irish et al., 1997], and other dynamic formations related to shallow-water sand movement. The oil, gas, and telecommunication industries are particularly interested in the use of lidar bathymetry for the determination of cable and pipeline landfalls in addition to shallow-water oil platforms. Previously uncharted or unknown locations of sea bottom objects, such as wrecks and obstructions, can also be determined by lidar bathymetry.

Rapid Response Environmental Assessments

A more recent application of lidar bathymetry is in the area of rapid response environmental assessments. Naturally occurring events, such as hurricanes, storm surges, and earthquake-induced tsunamis affect not only the physical geography of coastal areas, but also the economic, social, and human geography. Airborne lidar surveys prior to such events can be used to simulate many of these impacts through modelling and analysis [West and Wiggins, 2000]. The advantage of lidar bathymetry for post-disaster situations is its ability to deploy quickly, collect high volumes of data (day and night), and provide a fast result to first responders, scientists, and administrators. Hurricane Katrina struck many US states in the Gulf of Mexico in August, 2005. The US Army Corps of Engineers had been conducting bathymetric surveys in this area prior to the disaster, and then afterwards as well [Wozencraft and Lillycrop, 2006]. As the case study on shoreline change included in this chapter demonstrates, close estimates of the volume of change resulting from a natural disaster can be measured with lidar bathymetry.

Security and Defence

Modern society has developed an increased awareness of security, threat assessment, and military operations. Harbour protection, submarine hunting, and detection of mine-like objects are examples of primary concerns to governments at all levels. Lidar bathymetry plays a large role in assisting these efforts. From a military perspective, lidar bathymetry can help make tactical decisions for planning military operations such as shoreline assessments, amphibious assaults sites and mine-countermeasures [Lillycrop et al., 2000].

Environmental Monitoring

Although rapid deployment of lidar bathymetric surveys is beneficial for rapid response environmental assessments, environmental monitoring of the coastal zone is increasingly being seen as a priority for many government agencies. Monitoring the conditions of coastal shorelines through large regional airborne lidar bathymetry surveys is helpful in determining a benchmark of physical characteristics. As the debate surrounding rising sea levels ensues, it is important to map the potential for economic and physical damage and loss [Guenther, 2006]. Storms affect sandy coastlines more so than rocky ones, but flooding and wave surges are damaging to both. This can lead to erosion and shoreline instability. Coral reefs are particularly sensitive to environmental changes. Their growth, depletion, and health can in part be monitored by lidar bathymetry. This technology is non-invasive and allows for the mapping of fragile environments from the air. When combined with complementary technologies, such as hyperspectral imaging, aerial photography, multibeam sonar and side scan sonar, the information content increases significantly and enables investigators to gain a much broader understanding of the coastal zone.

Shoreline mapping is also a component of environment monitoring, as measurements of elevations both above and below the water line can change over time. The US Government is currently taking an inventory of the shoreline position and elevations of the 48 contiguous states, and plans to repeat the surveys with a temporal time span of approximately six years [Wozencraft and Millar, 2005, Wozencraft and Lillycrop, 2007]. This type of initiative will assist researchers in beach replenishment studies and erosion rates for susceptible areas.

River Surveying

Another interesting area that is gaining momentum in the lidar bathymetry community is surveys in a riverine environment. The goal of such surveys is not necessarily to map the river floor bathymetry, but rather to model shape and slope of the river system. Suspended sediment in the water column, extremely shallow water, and rough water due to increased flow volume or the presence of a steep gradient present limitations to river surveys. Some mitigating factors include undertaking the survey during reduced flow conditions and development of specialized algorithms to delineate the river bottom from the river surface. While river surveys have shown to be promising, there is not a large body of published research surrounding this topic. The 'Yakima River' case study in this chapter demonstrates one such example of a river survey.

Applications Summary

Hydrography, environment monitoring, rapid response environmental assessments, and security-related applications of lidar bathymetry are the primary roles for this technology. However, as airborne lidar bathymetry evolves and matures, new applications that are still being researched will continue to emerge into this ocean of opportunity.

CASE STUDY: COASTAL ENVIRONMENTAL MONITORING

Adapted from a previously unpublished paper
by original author Charles (Eddie) Wiggins
United States Army Corps of Engineers, Mobile, Alabama, USA

INTRODUCTION

The US Army Corps of Engineers (USACE) manages and executes its National Coastal Mapping Program (NCMP) through the Joint Airborne Lidar Bathymetry Technical Center of Expertise (JALBTCX). NCMP's mission is to map the shores of the United States on a recurring cycle, currently targeted between four to six years. The JALBTCX, a partnership between the Naval Oceanographic Office (NAVO), National Oceanic and Atmospheric Administration, and Corps of Engineers, utilizes the Compact Hydrographic Airborne Rapid Total Survey (CHARTS) system to collect topographic and bathymetric elevation data and imagery along the coastline [www.jalbtcx.org]. This seamless dataset provides foundational information for planning, engineering, constructing and operating Federal projects, and for assessment of regional sediment processes to support regional sediment management. The raw and derived information support the work of engineers, planners, economists, and environmentalists as they monitor and manage the nation's shores.

SURVEY HARDWARE

In 2003, JALBTCX took delivery of the CHARTS system from Optech. This sensor, owned by NAVO, combined a 9000 Hz topographic laser, a 1000 Hz hydrographic laser, and a 1 Hz RGB camera. In 2005, JALBTCX upgraded CHARTS to a 20000 Hz topographic and a 3000 Hz hydrographic laser and added an ITRES CASI-1500 hyperspectral imager.

The topographic sensor uses a 1064 nm wavelength laser and the hydrographic sensor produces coincident 532 nm and 1064 nm wavelength pulses. Through use of the green energy, the system measures depths to a maximum of two to three times the Secchi depth. Secchi depth is a term used to describe the clarity of waters. This measurement is made with a Secchi rod that is lowered into the water. Once the rod is no longer visible the depth is recorded defining the Secchi depth of the water being studied.

SURVEY SOFTWARE

Data collection, processing, and product generation require numerous software packages. SHOALS GCS, from Optech, provides the capability to generate survey flight mission plans and to process the raw data, producing points with calculated positions and elevations.

Fledermaus^{TM}, from Interactive Visualization Systems, allows point data editing in a three dimensional environment. A software package from NAVO, chartsLAS, writes point data files in ASCII and LAS formats. These point data files are the foundational information for grid elevation surface, shoreline vector, bare earth, and building footprint products created using a combination of Applied Imagery Quick Terrain Modeler®, ESRI ArcGIS®, and Visual Learning Systems Lidar Analyst®. A package from NAVO, pfm_charts_image, and a package from Leica, ERDAS Imagine® are used to process the 1 Hz images for creating ortho-mosaics. The programs, radcor and geocor from ITRES, TAAFKA from Navy Research Laboratory and ENVI® from ITT Corporation, process raw hyperspectral data to create fully-corrected reflectance hyperspectral mosaic images.

SHORELINE DATA COLLECTION OVERVIEW

The NCMP began in 2004 with surveys along the Gulf of Mexico and lower Atlantic shorelines. Data coverage extended along the northern Gulf of Mexico from Mississippi to Florida panhandle, the southwest Florida coast, and from Miami to North Carolina. The 2005 requirements covered from Virginia to Maine. Efforts in 2006 and 2007 focused on the Great Lakes, and Pacific coast surveys will follow in 2008 and beyond. Following the Pacific surveys, counter-clockwise coverage of the shorelines will begin repeat work in the Gulf of Mexico.

In addition to work conducted under the NCMP, JALBTCX often surveys stretches of coastline impacted by hurricane events. During the NCMP and storm response surveys in 2005 and 2006 JALBTCX collected over 6.5 billion data points and generated over 3 terabytes of lidar and imagery products. These regional-sized efforts focused on a 1500 m wide swath along the shore. Nominally, the survey covered 500 m landward and 1000 m seaward of the land-water interface. Table 1 provides the nominal parameters for CHARTS NCMP lidar collection efforts. The hyperspectral collection parameters produced 36 band images at 1-meter pixel resolution.

Because of the survey scope described above, valuable coastal data now exist for numerous regions of the US. More than 6.5 billion data points have covered over 4000 square miles of shoreline. As of November 2006, at least 1394 users from government, private, and academia sectors have downloaded over 18 billion elevations and 101 gigabytes of these data.

Table 2. CHARTS NCMP Data Collection Parameters

Topographic Data Collection	
Data Density	1 m
Swath Width	285 m
Planned Overlap	60%
Flight Ground-Speed	130 kts
Hydrographic Data Collection	
Data Density	4 m
Swath Width	300 m
Planned Overlap	30%
Flight Ground-Speed	180 kts

SHORELINE DATA APPLICATIONS

Due to the normal NCMP mission and the landfall of multiple hurricanes, JALBTCX surveyed the Alabama and Florida Panhandle coasts (Figure 11) five times within two years. The first survey in May 2004 provided pre-hurricane baseline information for the region. The following three collection efforts followed Hurricanes Ivan (2004), Dennis (2005), and Katrina (2005). The fifth survey (May, 2006) supported beach re-nourishment efforts. From these surveys came point clouds and other derived information. Because some of the efforts occurred shortly after major storm events, the data coverage varies because of differing water clarity.

Figure 11. Coastal Environment Monitoring Study Area.

Historically, scientists and engineers utilized cross section data to evaluate the condition of the coastal zone. These sections, taken at fixed locations and regular intervals, reveal the cross-shore differences between collection times. Figure 12 illustrates a cross section cut from CHARTS data collected near Panama City Beach, FL. This section demonstrates that the offshore and near-shore bars migrated closer to shore and that beach accreted at this location. However, cross sections do not reveal the conditions or changes between profile locations. High-density lidar provides valuable information covering virtually the entire region, and appropriate tools and techniques utilize the additional data.

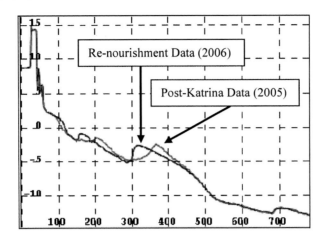

Figure 12. Elevation differences between re-nourishment and post-Katrina data.

Various products derived from lidar data support qualitative and quantitative data analysis. Surface differencing tools reveal significant changes in the terrain, both above and below water. Areas of erosion, accretion, development, and destruction appear quickly to the user. The one-meter digital elevation model in Figure 13 demonstrates differences between surveys acquired in May, 2006 and November, 2005. Areas of erosion and accretion are shaded respectively.

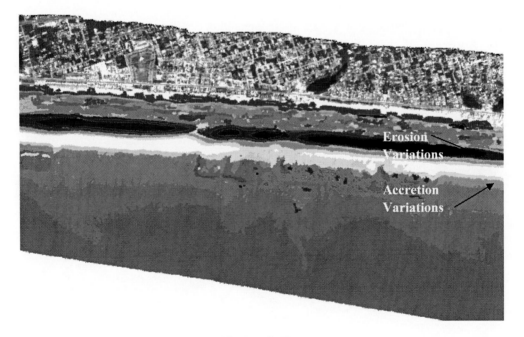

Figure 13. Elevation changes near Panama City Beach, FL.

The detail provided by the CHARTS data in the bathymetric DEM supports additional work aimed at determining the quantitative effects of the bottom topology on material movement.

Hurricane impact from Ivan (2004) and Katrina (2005) on a portion of Dauphin Island, AL was qualitatively determined using the bathymetric DEM data. The difference between the pre-Ivan and post-Ivan DEM data was visually displayed and compared to qualitatively assess the erosion and accretion impacts from this storm. Following Katrina the variations between the post-Ivan and post-Katrina DEM data were compared in the same manner. The comparison of these data allowed for the destruction of Katrina to be visually displayed and inspected. The locations of destroyed houses and buildings were determined and represented on the terrain for further analysis.

While qualitative assessment draws attention to trends and patterns, quantitative assessment of bottom change is necessary for engineering and design work. High-density bathymetric DEM information provides a more robust foundation for volume calculations than do traditional cross section or profile techniques. To facilitate the use of CHARTS NCMP data, the USACE Mobile District Spatial Data Branch developed a collection of enterprise GIS tools called eCoastal [http://ecoastal.usace.army.mil]. This enterprise GIS approach allows USACE customers around the country to uniformly perform common calculations and analyses.

Shoreline vectors calculated from a DEM quickly indicate areas of significant changes. Shoreline vectors generated from various surveys collected after Hurricanes Dennis and Katrina were displayed upon imagery in ESRI ArcGIS® (Figure 14). These vectors clearly define the alteration of the shoreline after each storm in the alongshore direction better than the cross sections. Further, eCoastal tools provide tools to calculate shoreline changes and rates of change from these vector locations.

Attributed building footprints were also generated in ESRI ArcGIS® from the lidar-based DEM. Attributes of these footprints include perimeter, area, and height above ground. Working in a GIS environment allows these vectors to be linked to propriety databases. After evaluating structural destruction caused by an event, economic impact calculations can occur more rapidly. Figure 14 demonstrates that building footprints and shoreline vectors extracted from CHARTS data provide an example of such calculations displayed upon concurrently displayed CHARTS imagery.

Figure 14. Building footprints and shoreline vectors

In 2004, engineers from multiple disciplines utilized the CHARTS coastal data and eCoastal tools to compute volumetric material changes throughout the hurricane impacted areas. These calculations were the basis of contract plans and specifications that resulted in more than $200 million of re-nourishment projects around the southeastern United States.

Volume computations also provide the foundation for regional sediment management practices as engineers and scientists identify and manage material transport patterns. Figure 15 portrays the Sediment Budget Analysis System for ArcGIS® using DEM data from Panama City Inlet, FL. The eCoastal tool provides a means for the user to define sediment volume changes within each region and allows capability to define the sediment transport quantities into and out of each box. This tool assists in regional sediment management activities.

CASE STUDY CONCLUSION

While products from CHARTS hyperspectral imagery have not yet matured, development is in progress. Environmental groups' efforts focus on generating habitat and resource maps. Data from subsequent surveys will allow calculation of rates of change. Various methods exist for depicting and analyzing survey data. The information required by the analyst should be the first consideration when deciding which tools to use for studying changes between different surveys. While cross sections may be the most familiar tool for the analyst, the surface DEM contains a significant amount of additional data that usually shows trends not easily seen in cross sections.

Figure 15. Sediment Budget Analysis System at Panama City Inlet, FL.

CASE STUDY: USING AIRBORNE LIDAR BATHYMETRY TO MAP SHALLOW RIVER ENVIRONMENTS

Adapted from a previously published paper by original authors David Millar[1], John Gerhard[2], and Robert Hilldale[3]
[1]Fugro Pelagos Inc., San Diego, California, USA
[2]Woolpert Inc., Denver, Colorado, USA
[3]Robert Hilldale, United States Bureau of Reclamation, Denver, Colorado

INTRODUCTION

Although Airborne Lidar Bathymetry (ALB) systems have existed for over 20 years the technology has rarely been used in riverine environments. Consequently, the "river survey" and the "shallow river survey," still present a major technical challenge to the surveyor.

In August 2004, the US Bureau of Reclamation, Woolpert, and Fugro Pelagos dispelled the myths and proved that ALB technology could in fact be used successfully to map shallow river environments. The Bureau of Reclamation required the creation of an accurate hydraulic model of the Yakima River in Washington State to study how changes in river flows would affect down stream fish habitat. Vessel-based acoustic and traditional land survey methods are both labor-intensive and do not yield full coverage. While ALB technology is very efficient and capable of providing full river bottom coverage, it is an industry accepted fact that ALB will not be successful in a shallow river environment.

After consultations, the Bureau of Reclamation decided to conduct an ALB survey during "low flow" conditions. The SHOALS-1000T ALB system was used to collect bathymetric soundings at a 2 m by 2 m sounding density on two reaches of the Yakima River. The maximum water depth was only 4.5 m. This project demonstrated how ALB technology can be used successfully in shallow-water river environments when survey risks are understood and managed.

Figure 16. Shallow river environment.

TECHNOLOGICAL / SURVEY BACKGROUND

For many years, river surveying has been carried out using vessel-based acoustic methods and/or traditional land survey methods. While reliable and accurate, both methods present significant limitations. They are time consuming and labor intensive. They can be dangerous to collect, especially during periods of high flow, and can exhibit significant site access concerns. Also, traditional techniques of surveying river environments do not provide full bottom coverage. In most cases, the river would be surveyed along a series of cross sections and profiles, requiring data interpolation when analyzing the final data set.

ALB has developed as a technology that addresses the limitations of vessel-based acoustic and traditional land survey methods. It is extremely efficient, with full coverage production rates of up to 70 km^2 per hour over large linear areas. Since it is an airborne remote sensing technology, there is no need for direct site access. The system can provide complete bottom coverage at densities up to 2 m by 2 m.

While ALB technology had been used with great success in "clear water" coastal environments for many years, it was always thought to have limitations in shallow-water river environments with poor water clarity. River systems transport sediments and often do not have the clarity required to support the ALB technology. The second major limitation regards bottom detection in shallow water, as it can be difficult to derive soundings. ALB systems have historically struggled with extracting returns in water depths of less than 50 cm. Algorithms had been developed to allow depth extraction in as little as 20 cm of water but were generally only effective in coastal environments. River environments often have large areas of shallow water, so such a limitation would result in significant areas where data could not be obtained.

Through a combination of new algorithm development and flow control, it was decided that ALB survey of the Yakima River in Washington State may be feasible. A survey attempt was completed with a pilot study area of two reaches of the Yakima River: the Easton Reach and the Kittitas Reach.

YAKIMA RIVER SURVEY

The survey was conducted in late August and early September, 2004. In an attempt to maximize water clarity, the flow on the river was controlled at the following dams: Cle Elum, Kachess, and Keechelus. During the acquisition time frame, the flow was restricted to 275 ft^3/s at the Easton Reach of the river and 3100 ft^3/s at the Kittitas Reach. The Bureau of Reclamation was interested in maintaining steady flow, as to not create additional suspended sediment. This allowed the survey team approximately five days to complete the ALB data collection.

The Bureau of Reclamation required a 2 m by 2 m sounding density and mission plans were developed based on this specification. A flying height of 300 m above the river was used, and the survey speed was 124 knots. These parameters required a 60 m swath width, resulting in several parallel lines to cover the extents of the meandering river.

The first reach surveyed was the Kittitas Reach requiring fifty flight lines and approximately 153 km to cover the entire survey area. The aerial survey took about four hours and covered approximately 5.8 km^2. The second reach surveyed was the Easton Reach. Forty-one flight lines and approximately 394 km were flown. The Easton aerial survey took five hours and covered approximately 17.8 km^2.

Positioning of the aircraft was provided using KGPS reference stations. One control point at each site was selected with two dual frequency GPS receivers set up at each control point during the aerial surveys. KGPS data were logged at a one-second interval, and this was combined with airborne GPS data to compute Post-Processed Kinematic (PPK) solutions of the aircraft trajectories. After the airborne survey was completed, all data were auto-processed in the field to verify complete project coverage.

During the airborne survey, both the Bureau of Reclamation and Fugro collected ground truth data separately at multiple locations in both of the survey sites. Their purpose was to check the results of the ALB survey. Since these data were collected independently from the air operation, the Bureau of Reclamation and Fugro were able to independently verify the integrity of the ALB data that was acquired.

DATA POST PROCESSING

The initial lidar post-processing demonstrated poor results. The processing algorithms were unable to derive soundings, despite the fact that the waveforms appeared fine. A trained operator could review the waveforms and see a very subtle bottom return. However, the post-processing algorithms were unable to "pick" the bottom returns displayed when reviewing the waveforms. The raw data were provided to the SHOALS-1000T manufacturer, Optech, and their algorithm experts. New algorithms were developed specifically to support ALB surveys in shallow-water river environments. Clear water and a highly-reflective river bottom are critical for determining bottom lidar returns. During the Yakima River survey the water was extremely clear due to the controlled flows. The bottom was also reflective, as it consisted of light-colored cobble.

The data were re-processed using the algorithms provided by the manufacturer. This processing iteration provided successfully derived soundings over most of both survey areas. The accuracy of these soundings was then established by comparing the ALB data to the GPS ground truth. In all cases, the ALB data matched the ground truth to within the quoted system accuracies. In fact, the accuracy was better than originally expected, even though some degradation was expected as a result of the new processing algorithms.

DATA ANALYSIS

After all of the data were delivered to the Bureau of Reclamation, the Bureau conducted their own independent analysis. In this case, the delivered ALB data were compared to the ground truth that was collected at the two survey sites. The analysis consisted of 14 cross sections and 4 longitudinal profiles created from the ground truth data. Two methods of comparison were evaluated to test the integrity of the data.

The first was a qualitative analysis of the ground-surveyed cross sections or profiles with the same data extracted from the ALB data. This was performed using a utility available for

ESRI ArcGIS®. Two separate TINs were generated for each site, one from the ground survey data and the other from the ALB data. Both were combined with the existing terrestrial lidar point data flown in November, 2000. The cross sections and profiles created from the ground-surveyed points were used to extract data from each TIN to conduct the analysis. The qualitative analysis proved that the ALB data and the survey points displayed similar accuracies.

The second evaluation method was statistical in nature and required obtaining the mean, median, and standard deviation of the vertical discrepancies between the ALB data and the ground survey. These data were organized in ArcMap by searching a radius of three feet from each ground survey point. The ground survey consisted of 328 data points, with 232 having at least one lidar data point within the 0.9 m radius. It was noted that by increasing the distance from a survey point, the elevation of the channel bed was more likely to vary from that of the survey point, resulting in an increased standard deviation.

The final data were classified to evaluate the lidar points falling within 0.3 m, 0.6 m and 0.9 m of a ground truth survey point. Due to the differences in quality of the ground survey between the Easton and Kittitas Reaches, the data were further classified to include all survey points Easton Reach points only and Kittitas Reach points only. Table 3 shows statistical information for the various breakdowns of the data analysis results. The values in Table 3 do not reflect any adjustments applied to the data.

Table 3. Elevation discrepancies between ground control and lidar data

	All Ground Survey Points			Easton Survey Points			Kittitas Survey Points		
Distance from ground survey point	<0.9m	<0.6m	<0.3m	<0.9m	<0.6m	<0.3m	<0.9m	<0.6m	<0.3m
Mean (cm)	0.16	0.16	0.11	0.13	0.13	0.11	0.21	0.19	0.11
Median (cm)	0.16	0.16	0.16	0.13	0.15	0.16	0.26	0.28	0.19
σ (cm)	0.30	0.31	0.24	0.22	0.24	0.18	0.39	0.39	0.35
Sample size	232	116	34	141	70	23	91	46	11

A data comparison of the Easton Reach, the shallower of the two reaches, exhibits that the standard deviation for the vertical measurements was within the stated system tolerance of 25 cm. Overall, the accuracy of the ALB appears to be biased high by approximately 15 cm. The comparison in the Kittitas Reach does not look as impressive, with a standard deviation of 39 cm. Again, there appears to be a bias for the data to display high results by an average of approximately 23 cm.

CASE STUDY CONCLUSION

Examining the data from the Easton Reach site shows that the accuracy and precision of the bathymetric lidar data displays similar accuracies to typical surveys conducted using sonar and GPS surveying equipment. The ALB survey data, although slightly biased high, is within acceptable bounds for hydraulic modeling, depending on the application. The ALB survey data from the Kittitas Reach appears to have a slightly higher bias and a larger standard deviation. The Bureau, Woolpert, and Fugro Pelagos have evaluated this data set and have come to the following conclusions:

1. There is a bias in many points of the Kittitas ground survey due to the survey rod not being perpendicular to the ground, thus biasing the ground survey low. The ground truth data from the Kittitas site were collected from a boat, since having a survey crew wade this reach was not possible. These collection techniques might be contributing to possibly inaccurate results.
2. The riverbed in both reaches contains cobbles, with Kittitas exhibiting slightly larger boulders. The ALB survey will generally pick the shoalest feature within the ALB footprint. Unless the ground survey also picked the shoalest feature, this will bias the ALB data high.
3. The sample sizes used for the initial accuracy analysis were relatively small for both reaches, with 141 points available for comparison on the Easton Reach and only 91 points available for comparison on the Kittitas Reach. A much larger sample size would be required to draw reasonable and accurate conclusions.

All parties acknowledge that further investigations are required. Much more ground survey data will be collected during the next aerial survey. Since the initial ALB data collection in August 2004, five more reaches of the Yakima River were flown for ALB data in April 2005. A more dense set of ground truth was collected in these reaches at the time of the ALB data collection. A more in-depth analysis of the accuracy and precision obtained during this data collection and subsequent flights covering other reaches will be submitted to an appropriate journal in the coming months.

The density of the point data from an ALB survey is a vast improvement over sonar or GPS surveys. In fact, one can argue that even if there is a slight degradation in accuracy, the improved coverage provides a better representation of the channel bed. Furthermore, the costs of an ALB survey are similar to those of a sonar data collection by boat when the reach is greater than 24-32 km. The larger the project, the more feasible ALB becomes. However, this does not mean that shorter reaches should not take advantage of the ALB approach. Savings can be realized through reduced mobilization and demobilization costs if the survey can be coordinated with other survey efforts in the same region.

In summary, it is important to consider the project requirements with respect to resolution, accuracy, and precision before performing any type of survey. For many hydraulic studies, ALB surveys do meet these requirements. The expansion of existing technology will greatly improve the ability of the hydraulic engineer to model longer reaches of a river with a reasonably accurate representation of the riverbed. It is anticipated that with further improvements in software and hardware, resolution, accuracy, and precision of ALB surveys will also improve.

SUMMARY

Lidar bathymetry is still emerging as a technology that will surely grow in application and use in the coming decades. Many organizations are turning to lidar bathymetry for surveying the sea floor in shallow water environments because it is versatile, accurate, and comparatively inexpensive. By employing two wavelengths of the electromagnetic spectrum, infrared and green, sea floor depths in clear waters generally less than 70 m can be accurately measured from either fixed-wing or rotary-wing aircraft. From inception in the early 1970s, lidar bathymetry has evolved from a single pulse, single laser system to a dynamic laser, scanning technology that is tightly integrated with motion sensors, GPS, imaging cameras, and topographic lidar. Many products are derived from lidar bathymetry, including digital elevation models, sea floor reflectance images, reduced sounding sets, TINs, contours, and other digital products. These products help drive the growing list of applications that use lidar bathymetry. Although navigational hydrographic applications are still being verified by many agencies, interest in this domain is growing as the technology matures and accuracy levels for small bottom objects and bottom sensing in turbid waters improves. Non-navigational critical applications, such as coastal zone monitoring, security, and defence and river surveying are all expanding the need for this technology.

As lidar bathymetry continues to evolve, the benefits will be clear. Improved laser repetition rates and narrower beams of light will enhance the technology. New lightweight systems and compact systems that are commercially produced will assist in driving costs down and availability up. Hardware and software improvements increase both processing times and accuracy of measured data. The value of collected information will boost and expand with better bottom reflectance classification and water column characteristics captured by volume backscatter. Finally, sensor fusion will significantly enhance the data value content through integrated technology such as hyperspectral imagers, additional lidar transmitters/receivers, and digital imagery and video.

Although bathymetric lidar has been around for many years, it is now emerging as a significant asset to the remote sensing community and geospatial data providers and consumers.

ACKNOWLEDGMENTS

The authors wish to thank those who contributed to this chapter on lidar bathymetry. Each of the three main commercial vendors of lidar bathymeters was helpful in providing experiences, anecdotal information, papers, figures and technical knowledge. We are indebted to Andreas Axelsson, of Airborne Hydrography AB (Sweden) and Rhys Barker of Tenix Inc. (Australia) for their assistance and help with technical information and case studies. David Millar and Carol Lockhart of Fugro-Pelagos (USA) were very helpful in providing us case study content and resources on data processing. Several people at Optech Ltd. (Canada) were kind enough to assist us with figure and image materials, technical content and time. The US Army Corps of Engineers and their wealth of experience and knowledge was particularly beneficial to us and we thank Jeff Lillycrop, Jennifer Wozencraft, and Charles (Eddie) Wiggins for their patience, support, case study and information sharing. Our academic and editorial reviewers were critical to the success of this chapter, and we appreciate the talents of Tamrat Belayah, David Millar, Fred Persi, Kimberly Kearns, Andrew Weitz, Jennifer Wozencraft and Julie Wright. We are also very grateful to ESRI Inc. for its impartial support in writing this chapter. Two very key people who enabled the authors to work on the chapter, provided support at their own personal expense and continue to help us grow as professionals cannot be left out: we sincerely appreciate Andrew Weitz and Kimberly Kearns.

REFERENCES

Bonham-Carter G.F. (1995). *Geographic Information Systems for Geoscientists: Modelling with GIS*, Pergamon Press, New York, USA.

Clegg, J.E. & Penny, M.F. (1978). Depth Sounding From Air By Laser Beam, *Journal of Navigation*, Vol. 31, p. 52.

Danson, E. 2006). *Understanding Lidar Bathymetry for Shallow Waters and Coastal Mapping*, Shaping the Change XXIII FIG Congress, Munich, Germany, October 8-13.

Elachi, C. (1987). Introduction to the Physics and Techniques of Remote Sensing, John Wiley & Sons, Inc., USA, pp 22-23.

Engstrom, R. & Axelsson, R. (2001). Laser bathymetry and its compliance with IHO S44, Proc. Hydro 2001, *The Hydrographic Society, Special Pub*. 42, March 27-29, Norwich, England, Paper 18, pp. 1.

Guenther G.C. (2006). Chapter 8: Airborne Lidar Bathymetry, Digital Elevation Model Technologies and Applications: *The DEM Users Manual*, American Society for Photogrammetry and Remote Sensing, Maryland, USA, pp. 237-287.

Guenther, G.C., Cunningham A.G., LaRocque, P.E., et al. (2000). Meeting the accuracy challenge in airborne Lidar bathymetry, *Proc. 20th EARSeL Symposium: Workshop on Lidar Remote Sensing of Land and Sea*, European Association of Remote Sensing Laboratories, Dresden, Germany, pp. 23.

Guenther G.C. & Mesick, H.C. (1988). Analysis of Airborne Laser Hydrography Waveforms, *Proc. SPIE Ocean Optics* IX, v925, pp. 232-241.

Guenther, G.C., LaRocque, P.E. & Lillycrop, W.J. (1994). Multiple Surface Channels in SHOALS Airborne Lidar, *Proc. SPIE Ocean Optics* XII, Vol. 2258, pp. 422-430.

Guenther, G.C. & Goodman, L.R. (1978). "Laser Applications for near-shore nautical charting", *Proc. SPIE Ocean Optics* V, v160, 1978, pp.174-183.

Guenther, G.C. (1986). Wind and Nadir Angle Effects On Airborne Lidar Water Surface Returns, *Proc. SPIE Ocean Optics* VIII, v637, pp. 277-286.

Hickman, G.D. & Hogg, J.E. (1969). Application of an Airborne Pulsed Laser for Near-Shore Bathymetric Measurements", *Remote Sensing oOf Environment* 1, p. 47.

Houck, M.W. (1980). The Hydrographic Airborne Laser Sounder (HALS) Development Program, *Proc. Laser Hydrography Symposium*, Australian Defence Research Centre, Salisbury, South Australia, p. 35.

Irish, J.L. & Lillycrop, W.J. (1997). Monitoring New Pass, Florida with High Density Lidar Bathymetry, *Journal of Coastal Research,* v13, No. 4, pp. 1130-1140.

Irish, J.L., Lillycrop, W.J. & Parson, L.E. (1997). Accuracy of Sand Volumes as a Function of Survey Density, *Proc. 25th International Conference on Coastal Engineering, American Society of Civil Engineers*, September 2-6, Orlando, USA, v3, pp. 3736-3749.

Irish, J.L., Truitt, C.L. & Lillycrop, W.J. (1997). Using High-resolution Bathymetry to Determine Sediment Budgets: New Pass, Florida, *Proc. 1997 National Conference on Beach Preservation Technology, Florida Shore and Beach Preservation Association*, pp. 183-198.

Kletzli, R.G. & Peterson, J.L. (1998). Inertial Measurement and Lidar Meet Digital Ortho Photography: A Sensor Fusion Boon for GIS, Proc. *1998 ESRI GIS User Conference*, San Diego.

Levis, C.A., Swarner, W.G., Prettyman, C., et al. (1973). An optical radar for airborne use over natural waters, *Proc. Oceans '73*, pp. 76-83.

LaRocque P.E. & West, G.R. (1999). Airborne laser hydrography: an introduction, *Proc. ROPME/PERSGA/IHB Workshop on Hydrographic Activities in the ROPME Sea area and Red Sea*, October 24-27, Kuwait, pp. 16.

Lillycrop, W.J., Irish, J.L., Pope, R.W., et al. (2000). GPS Sends In The Marines: Rapid Environmental Assessment With Lidar, *GPS World*, 11, 11 November, pp. 18-28.

Ott, L.M. (1965). Underwater Ranging Measurements Using Green Laser, *NAVAIRDEVCEN Report No. NADC-AE-6519*, Naval Air Development Center, Warminister, Pa. (Confidential).

Sorenson, G.P., Honey, R.C. & Payne, J.R. (1966). *Analysis of the use of airborne laser radar for submarine detection and ranging*, SRI Report No. 5583, Stanford Research Institute.

Thomas, R.W.L. & Guenther, G.C. (1990). Water Surface Detection Strategy for An Airborne Laser Bathymeter, Proc. Ocean Optics X, SPIE, v1302, pp. 597-611.

Tuell, G., Park, J.Y., Aitken, J., et al. (2005). Adding Hyperspectral to CHARTS: Early Results, Proc. U.S. Hydro, The Hydrographic Society of America, San Diego, USA, p. 10.

West, G.R. & Wiggins, C.E. (2000). Airborne Mapping Sheds Light on Hawaiian Coasts And Harbors, *EOM (Earth Observation Magazine)*, April, v9, No. 4, pp. 25-27.

Wozencraft, J. & Millar, D. (2005). Airborne Lidar and Integrated Technologies for Coastal Mapping and Nautical Charting, *Marine Technology Society Journal*, Autumn, v39, No. 3, pp 27-39.

Wozencraft, J.M. & Lillycrop, W.J. (2006). "JALBTCX Coastal Mapping for the USACE," *International Hydrographic Review* 7(2), pp. 28-37.

Wozencraft, J.M., Macon, C.L. & and Lillycrop, W.J. (2007). "CHARTS-enabled data fusion for coastal zone characterization," Proceedings Coastal Sediments 2007, ASCE in press.

Wright, C.W., Brock, J. (2002). EAARL: A Lidar for Mapping Shallow Coral Reefs and Other Coastal Environments, *Proc. Seventh International Conference on Remote Sensing for Marine and Coastal Environments*, ERIM, Miami, USA.

In: Laser Applications in Environmental Monitoring
Editors: L. Fiorani and F. Colao

ISBN 978-1-60456-249-1
© 2008 Nova Science Publishers, Inc.

Chapter 2.2

LASER INDUCED FLUORESCENCE

A. Palucci

Laser Applications Section, Advanced Physical Technologies and New Materials
Department, Italian National Agency for New Technologies, Energy
and the Environment (ENEA),
via Enrico Fermi 45, 00044 Frascati, Italy
E-mail: antonio.palucci@frascati.enea.it, tel.: +39 – 06-94 00 52 99

ABSTRACT

Thanks to recent technological advances, the Laser Induced Fluorosensor (LIF) systems are now emerging from the research fields of the scientific community and by virtue of their capabilities migrate toward an impressive range of applications spanning civil and industrial.

This paper will deal with the physical background of LIF technique, as well as with a detailed description of several optical emission bands which are relevant for practical applications. The main components of a lidar fluorosensor apparatus will be presented as well, within the different operational modes (single or dual pulse emission). Application of LIF systems will be given with a particular emphasis on hydrographic measurements.

INTRODUCTION

The development of investigation tools for identification, mapping and evaluation of natural or pollutant components is a key action in environment protection, ecosystem safeguard and life sustaining. Remote operation, prompt response, real-time analysis, and large area scanning are the demands and the skills requested in order to supply and establish guidelines for conservation, management and remediation actions.

Laser Induced Fluorescence (LIF) can be very effective in significant measurements of the bio-optical parameters in vegetation and/or natural waters. This laser technique has proved popular [1] due to its exceptional sensitivity and low mass detection limits [2]. Qualitative (flow visualization) and analytical applications include measurements of gas-phase concentrations in the atmosphere [3], plant health status [4], in monitoring water bodies

[5] and crude oil pollutants releases [6], in experimental fluid mechanics to measure the concentration of a scalar species within a fluid [7], for fuel visualization in engine environments [8], and in determining the density of a certain atomic level in the plasma directly from the absorption coefficient [9].

LIF is the optical emission from atoms or molecules that have been excited to higher energy levels by absorption of electromagnetic radiation. The main advantage of fluorescence detection compared to absorption measurements is the greater sensitivity achievable because the fluorescence signal has a very low background. In the case of resonant excited molecules, LIF provides selective analyte excitation, thus avoiding interferences. LIF can be profitably applied to study the electronic structure of molecules and to perform quantitative measurements of analyte concentrations.

LIF technique employs electro-optics components and can therefore gain the advantages of their peculiarities, such as in setting-up diagnostics tools, to be local and/or remote, in real time, and to operate with low radiation exposure so that the sampled object is not disturbed or damaged at all. In detail, LIF offers the advantages of being:

- fast (the detection of a substance can be performed in a fraction of a second);
- remote (the system and the target can be meters apart);
- sensitive (better than parts-per-million);
- specific (substances can be recognized by their spectroscopic fingerprint);
- user-friendly (the system can be deployed in few minutes and does not require a specifically trained user).

Based the above considerations, it's straightforward to include LIF in laser application techniques for environmental monitoring, commonly called Lidar Fluorosensor. Different platforms were employed to install the monitoring apparatus, ranging from fixed installations [10], mobile platforms [11, 12], airborne [13, 14, 15], ship vessels [16, 17, 18] and submarine payloads [19, 20], both for vegetation and water remote sensing. We will restrict our interest to marine applications of this technique.

OPTICAL EMISSION

As it is common for lidar systems, a fluorosensor apparatus employs a laser source to transmit coherent monochromatic photons towards the target; then the backscattered radiation is collected and analyzed by an appropriate optical system [21].

While in elastic scattering the main contribution can be ascribed to Raleigh and Mie scattering, inelastic scattering occurs when a light photon is absorbed and re-emitted in a different wavelength range from the excitation. In this respect, Raman and Brillouin scattering, as well as fluorescence by specific chromophores, are forms of inelastic scattering. A chromophore is part (or moiety) of a molecule responsible for its color and generally embedded in dissolved or suspended molecules as CDOM (Chromophoric Dissolved Organic Matter) or phytoplankton pigments (chlorophyll-a, carotenoids, phycocyanine, phycoerythrin).

Combined measurements of elastic and inelastic scattering, as well as fluorescence and reflection, may allow simultaneous definition of location of the target, its size and chemical composition, as in laser cytometry application of phytoplankton analysis.

In the case of lidar fluorosensor, the main contributions result from (1) laser backscattered radiation (i.e., in the case of water surface specular reflection) and, depending on the peculiarities of the target materials excited by the laser light, (2) frequency-shifted emissions (i.e., inelastic scattering and fluorescence) as schematically shown in Figure 1.

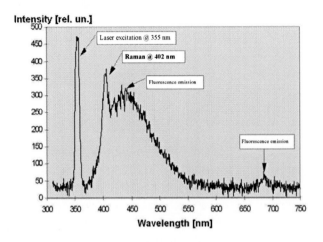

Figure 1. Optical emission emitted by seawater sample, upon UV laser excitation (@355 nm). From the left: the laser specular reflection, the inelastic water Raman scattering and two broad fluorescing contributions, respectively.

The reflection at the target surface can generally be avoided by employing a suitable band-pass filter. On the other hand, the monochromatic excitation can produce a wide wavelength range emission, carrying specific information about the chemistry or the physical state of the target material. After laser excitation, photons are absorbed and part of the energy associated with them can, through internal relaxation mechanisms, be reemitted as radiation at a longer wavelength than the original absorbed light (Figure 2).

In the case of homogeneous seawater, assuming a linear regime for the laser excitation and low chromophore densities for all the present species (natural offshore seawater) saturation can be neglected. The space integration on the investigated water column generates a total time integrated LIF signal $F(\lambda_{em})$, which can be expressed as [20]:

$$F(\lambda_{em}) = \frac{A(R_0)}{m^2} E_{ex} \frac{\sigma(\lambda_{em}, \lambda_{ex})}{k_T} \qquad (1)$$

where λ_{em} (λ_{ex}) stands for the emission (excitation) wavelength, m is the refraction index of water, $A(R_0)$ is a constant embedding most of the above mentioned system parameters and changing with the distance R_0 from the water surface, E_{ex} is the excitation pulse laser energy, σ is the fluorescence efficiency of the process, and $k_T = k_{ex} + k_{em}$ is the total extinction coefficient resulting from extinction terms at the excitation and emission wavelength.

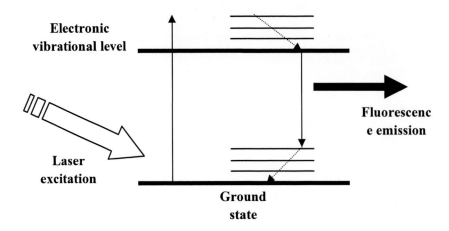

Figure 2. Schematic diagram of laser excitation and fluorescence emission.

Raman scattering is a very important lidar fluorosensor feature, because its contribution to the received spectrum is highly molecule specific, and its intensity is directly proportional to the concentration of the Raman emitting material. In inelastic scattering the absorbed laser radiation raises the energy of the molecule to an excited level from which it immediately decays (in $<10^{-14}$ s) to its ground state with concurrent emission of radiation that has a wavelength different than that of the excitation light source. The difference in energy between the incident and emitted photon is a characteristic of the irradiated molecules. The intensity of Raman scattering signal is inversely proportional to laser excitation wavelength, while the frequency shift (also known as Raman shift) is determined by vibrational and rotational transitions in the molecule. In the case of water, 3400 cm^{-1} is the frequency shift for the OH stretch vibrational mode [22].

Moreover, in order to evaluate concentrations of different chromophores dispersed in water, LIF signals can be calibrated against the concurrent water Raman signal, regarded as an internal standard reference [23]. By rationing the chromophore fluorescence signal F to the water Raman intensity R, we have:

$$\frac{F}{R} = \frac{\sigma_F}{\sigma_R} \frac{k_{ex} + k_R}{k_{ex} + k_F} \tag{2}$$

where the indexes are self explanatory and the dependence on system parameters and on the refraction index of water has disappeared.

The ratio of extinction coefficients in Equation (2) can approximately be regarded as a constant and thus neglected, provided a careful choice of excitation and emission wavelengths is performed, in order to avoid errors due to differential absorption. In conclusion, the different chromophore concentrations, expressed in Raman units, are seen to be independent of system parameters. This procedure is usually followed by proper calibration with the matter investigated.

In general, the experimental system is located above the sea water (Figure 3), at a range R_w from its surface, and the laser beam, after propagation in air, probes a water layer characterized by the extinction coefficient α_w.

The choice of the laser excitation wavelength and the operational height of the lidar fluorosensor apparatus is a trade-off between the extinction coefficients, in air and water, and the excitation efficiency. Even if many substances present, for example, in the sea water are more efficiently excited in the UV wavelength range, a lower limit can be observed downwards 300 nm, due to the strong absorption of shorter wavelengths both from the atmosphere and water. For this reason, and for eye-safety considerations as well, the most common excitation wavelength lies in the near UV, at 308 nm in the case of excimer gas laser or at 355nm in case of solid state Nd:YAG laser with third harmonic generation.

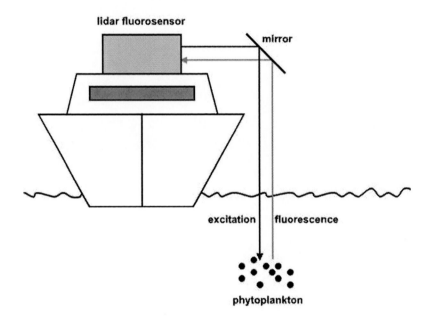

Figure 3. Lidar fluorosensor principle of operation.

THE LIDAR FLUOROSENSOR APPARATUS

A typical fluorosensor instrument is quite similar to a more general lidar layout, with a near UV or visible laser transmitter, sending/receiving optics and control electronics. The difference relies on the signal discrimination, generally performed with the help of a dispersive elements (grating, interference filters) behind the optical detectors (photodiode array, photomultiplier tubes).

As a reference, the main characteristics of the ENEA Lidar Fluorosensor (ELF) [17] are reviewed and some recent results are described. ELF has been operational aboard the research vessel (RV) *Italica* during four Italian expeditions in Antarctica (13th, 15th, 16th and 18th in 1997-98, 2000, 2001 and 2003, respectively) [24] and the MIPOT (Mediterranean Sea, Indian and Pacific Oceans Transect) oceanographic campaign (2001-02) [25] As we will see in the next section, its data have been used for the validation and/or the calibration of space borne radiometers [26, 27], up to the calculation of new estimations of satellite sensed primary production (PP) [28, 29] and CDOM.

ELF (Figure 4) is part of a complete laboratory, including local and remote instruments for continuous monitoring and in situ sampling, lodged into an ISO 20' container. It is assisted by ancillary instruments: a lamp spectrofluorometer, a pulsed amplitude fluorometer (PAM), a solar radiance detector, measuring the photosynthetically available radiation (PAR), and a global positioning system (GPS).

Figure 4. Side view of ENEA Lidar Fluorosensor (a front view is given in the upper left insert). The transmitter (1) is a frequency-tripled Nd:YAG laser; the receiver (2) is a Cassegrain telescope; the detection sub-system is composed by a four-arm optical fiber bundle (3), interferential filters (4) and photomultipliers (5).

The light source is a frequency-tripled Nd:YAG laser (355 nm) followed by a beam expander (BE); transmitter and receiver are mounted on a common chassis with coincident axes, in order to minimize the possible optical mismatches induced by mechanical vibrations and shocks. The laser beam is expanded by a factor of three before reaching the sea, both for eye-safety restrictions and in order to increase the laser footprint. The telescope collects and focuses the return optical radiation onto a fiber optic front tip, which is placed behind a high-pass optical filter used to cut off the laser radiation elastically backscattered. This four-arm fiber optic bundle splits the collected fluorescence light, routing the signals to interferential filter (IF) to select spectral band and finally to four different photomultipliers. The electronic analog output is digitized by charge sensitive Analog-to-Digital Converters. A personal computer embedded in a Versa Module Eurocard bus controls all the experimental settings, including the triggering schema for excitation (normal and/or pump-and-probe), the laser transmitter energy, the photomultipliers high voltage and gating time and the data acquisition parameters. The lidar fluorosensor arrangement on-board the ship is schematically sketched in Figure 4, while detailed characteristics of the main components are listed in Table 1.

Table 1 - Typical specifications of ELF

Transmitter	Laser	Nd:YAG
	Wavelength	355 nm
	Pump pulse energy	30 mJ
	Probe pulse energy	3 mJ
	Pulse duration	10 ns
	Pulse repetition rate	10 Hz
	Beam expansion	$1 \times - 20 \times$
Receiver	Telescope	Cassegrain
	Clear aperture	0.4 m
	Focal length	1.65 m
	OF length	1 m
	OF diameters	Input 25.4 mm, output 7 mm
	IFs center wavelength	403, 450, 580 (or 650), 685 nm
	IFs bandwidth	5 nm FWHM
	PMs	Hamamatsu R3896, R1477, R928
	Gated HV	$100 - 200$ ns
Electronics	Bus	ISA[a]-VME mixed bus
	ADC	CAEN V265 (15 bit)
	PC	Embedded CPU[b] Intel 486, 100 MHz

[a] Industry standard architecture.
[b] Central processing unit.

Figure 5. Main picture: RV *Italica*; ELF and the ancillary instruments are housed in a container (inside the circle). Left insert: external view of the container; note on the left of the container the box accommodating the mirror for water surface observation. Right insert: internal view of the container; ELF and the spectrofluorometer are visible behind and on the left of the operator, respectively.

By placing suitable narrow band interference filters in front of the each photomultiplier, four spectral channels were selected, corresponding to the water Raman backscattering (403 nm), the DOM fluorescence maximum (450 nm), the DOM red tail emission (650 nm) and the Chlorophyll peak (685 nm), respectively. The choice of the latter two channels relies on literature knowledge of the prevalent phytoplankton composition. In the case of

Chlorophyceae, characterized by a dominant chlorophyll fluorescence emission between 680-690 nm this has been held responsible for at least 90% of the biomass of the Indian Ocean in the absence of algal species characterized by the presence of other pigments emitting on different red spectral regions (phycoerythrin @ 580 and phycocyanin @ 650 nm).

The spectral and electrical responses of the photomultipliers and the related electronics are tested before and after the campaign by using standard lamps and fluorescing targets during the calibration procedure.

SEAWATER FLUORESCING MATTER

As the aerosols in atmosphere, the marine hydrosols are made by different semi-liquid or solid components floating in seawater, mainly in inorganic and organic classes. The last ones can further be separated into Particulate Organic Matter (POM; i.e., phytoplankton) and Dissolved Organic Matter (DOM). POM is the residual material left on a 0.45µm filter paper, after seawater filtration. Particulate Organic Carbon (POC) is the total carbon fraction of POM. POC is dominated by detritus-like material ($\sim$0.03 10^{18} gC), while living biomass is just a minor component ($\sim$0.5-1.0 10^{15} gC). Planktons are among the most important components of POC in the photic zone.

The organic matter can be classified from its source as:

Allochthonous
- Atmosphere
- Rivers/estuaries/coastal zones

Autochthonous
- Photosynthesis (living organisms)
- Detritus
    - dead organisms
    - fecal material
    - organic aggregates
    - flocculation

The main POM components are listed in Table 2.

Table 2 – Detail of the POM components and relative origin

Components	Source
Proteins (amino acids)	Auto and Allo
Carbohydrates	Auto and Allo
Lipids	Auto and Allo
Pigments	Auto and Allo
Lignin	Allo
Nucleic Acids	Auto and Allo
Other Metabolites	Auto and Allo

Phytoplankton is composed of a mixture of microscopic plants that live in the oceans as well as in different aquatic environments and can mostly be considered a good indicator of the biogeochemical oceanic cycles because its growth depends on stable and favorable surrounding conditions. It is the base of the trophic sea chain, and produces almost all of the oxygen released by the oceans in the atmosphere, being also responsible for over 40% of the global carbon fixation. Phytoplankton biomass can be estimated by the chlorophyll-a concentration. Marine ecosystem supports more than 35% of the total (marine and terrestrial) primary production with a very fast turnover, weeks and more than ten years, respectively [30]. It can be considered the grass of the sea [31] and as a consequence of its internal composition radically influences the color of the oceans, also affecting the depth of light penetration through the water column [32].

Many phytoplanktonic species are able to control their vertical position in the water column, but can float following the marine currents or by flagellar mobility. Their migration from surface to sub-surface layers are mainly due to variations in nutrients and radiance availability along the water column.

Seven main taxonomic groups of phytoplankton can be recognised [33, 34, 35]: *Chlorophyta* (Green Algae); *Cyanophyta* (Cyanobacteria or Blue-green Algae); *Bacillariophyta* (Diatoms); *Chrysophyta* (Golden-brown Algae); *Cryptophyta* (Cryptomonads); *Pyrrhophyta* (Dinoflagellates) and *Euglenophyta* (Euglenoids). As an example, SEM (left) and fluorescence (right) microscope images of *Dinophyceae Prorocentrum minimum* are shown in Figure 6.

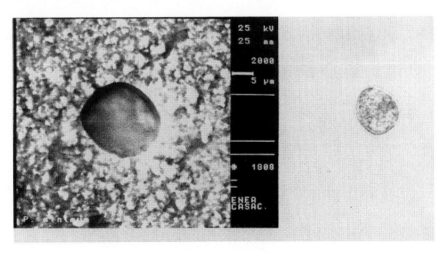

Figure 6. SEM (left) and fluorescence (right) microscope images of *Dinophyceae Prorocentrum minimum.*

Algal pigments consist of chlorophyll a, b and c, β-carotene, some xanthophylls, phycobilins, phycocyanin, and phycoerythrin. Size is one of the main characteristics of these unicellular compounds that can be included in the middle between the very small viruses and larger zooplankton species. Micro-planktons are the main components of the natural algal community, while pico-plankton contribution is not negligible.

FLUORESCENCE YIELD MEASUREMENT BY PUMP AND PROBE LIF

The algal photosynthetic apparatus allows for light energy conversion into photochemistry processes as carbon dioxide and water are converted into organic molecules. Light harvesting antennas (Chlorophyll-a) are employed to absorb and collect the blue and red radiation and trigger the Calvin cycle. The main processes involved in photosynthesis are photochemistry, non photochemical quenching and fluorescence.

The fluorescence detection technique is especially suitable for investigating the photosynthesis process in living vegetation tissue, having the advantage of being nonintrusive and in supplying real-time information either in laboratory or in field experiments [4] revealing possible stress status (nutrient depletion, presence of toxicants, photo-inhibition, etc.). A double pulse laser remote sensing system has been developed to be operated according to the Pump and Probe (PP) LIF technique [36] in order to directly measure the algal fluorescence quantum yield.

In the application of the PP technique to the vegetation monitoring, chlorophyll LIF signals around the 690 nm band (labeled F_i) are used to gain information about the investigated target photosynthetic efficiency. An isolated probe laser pulse in the dark or in the presence of background illumination triggers a fluorescence signal (F_0 or F_s) corresponding to the actual partial closure of the Reaction Centres (RCs). An intervening intense laser pulse (pump) switches all the RCs to the closed state, of which the fluorescence response (F'_m) can be tested by a second probe pulse, provided that the pump-to-probe delay is kept within the quenchers' lifetime, which is of the order of 100 μs. The conditions necessary to fully close the RCs and to leave them unperturbed when probed have been studied in detail in Refs. [36, 37].

The fluorescence quantum yield Y_{pp} can be computed as:

$$Y_{PP} = \frac{\Delta F}{F'_m} = \frac{F'_m - F_s}{F'_m} \tag{3a}$$

which at dark corresponds to:

$$Y_{PP} = \frac{\Delta F}{F_m} = \frac{F_m - F_0}{F_m} \tag{3b}$$

The chlorophyll signal intensity F_i, used in Equation (3) are those obtained from the rough data after proper normalization procedure. Whereas the intrapulse normalization (Equation 2) is necessary to obtain chlorophyll concentration in Raman units, different interpulse normalization, performed on the same spectral channel measured in different LIF spectra, is required to account for the specific change in the efficiency of in-vivo chlorophyll fluorescence. In the latter case, it is required to measure also the excitation laser energy which must be kept in the final expression utilized for the pump-and-probe data.

The interpulse normalization requires the rationing of the fluorescence signal on the respective laser energy, according to the following:

$$\frac{F_\lambda}{E_{ex}}\bigg|_p = \frac{A}{m^2}\frac{\sigma_{P1}}{k_T} \tag{4a}$$

$$\frac{F_\lambda}{E_{ex}}\bigg|_{PP} = \frac{A}{m^2}\frac{\sigma_{P2}}{k_T} \tag{4b}$$

where F_λ in Equation (4a) is the fluorescence signal induced by the probe pulse alone (P1) with a cross section σ_{P1}, relevant to the unperturbed RCs. While F_λ in Equation (4b) is the fluorescence signal induced by the probe pulse (P2), after that the RCs have been closed by the pump with a cross section σ_{P2}, according to the time sequence sketched in Figure 7. Namely, remotely sensed data are taken detecting the LIF signals induced by probe pulses P1 and P2, where the optimum P2 to pump delay is 50 μs, as experimentally determined [37].

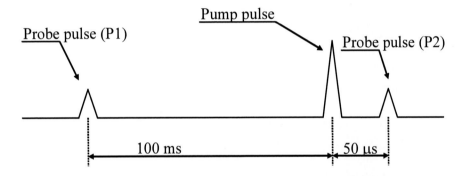

Figure 7. Time sequence of laser pulses used in the PP technique: probe pulse P1 is used to test the pre-saturation fluorescence status; pulse P2 is sent 50 μs after the pump pulse, to stimulate the maximal fluorescence.

By introducing Equation (4a) and (4b) into (3a) and (3b) we obtain:

$$Y_{PP} = \frac{\sigma'_m - \sigma_s}{\sigma'_m} \tag{5a}$$

and

$$Y_{PP} = \frac{\sigma_m - \sigma_0}{\sigma_m} \tag{5b}$$

for a photosynthesizing target exposed to light- and dark-adapted, respectively.

Due to this normalization all the instrumental factors are removed, thus resulting in an absolute value of the fluorescence quantum yield.

The contemporary measurement of PAR and laser fluorescence yield (Y_{PP}) allows obtaining information on the ETR (Electron Transport Rate), that represents the apparent photosynthetic electron transport rate in μmoles electrons m^{-2} s^{-1}, which can be calculated on the basis of the measured values of Y_{PP} and PAR using the equation:

$$ETR = Y_{pp} * PAR * 0.5 * k \qquad\qquad (6)$$

where the factor 0.5 considers the absorption of two quanta for the transport of one electron and the constant k includes the collecting optical efficiency of the harvesting antennas. In the case of phytoplankton, k value is generally assumed to be close to 1 while a 10% of absorption can be assumed due to the extinction coefficient of water at 355 nm in the first few layers, giving rise to $k=0.9$, used in the present data analysis.

Laser pump-and-probe experimental tests, with excitation @ 355 nm or 532 nm, were used to detect the presence of toxicants, and the effects of plant exposure to thermal stresses and to low levels of gaseous pollutants [37]. Laser measured fluorescence yields (Y) have been already compared with those obtained by an in-situ fluorometer (PAM), and laboratory experiments have shown that the appropriate choice of experimental parameters (pump and probe laser intensities), make Y to approach the theoretical value expected for healthy dark-adapted targets.

In concluding this section, it is important to remind the reader that in estimating the biomass from LIF measurement, an important and often underestimated effect is related to the change in fluorescence intensity emission due to environmental parameters and mainly from the sun level light. Indeed, the chlorophyll-a fluorescence F(680) depends on the photosynthetically active radiation PAR that influences the chlorophyll-a reaction centers:

$$F(680) = \frac{F_{raw}(680)}{1 - \dfrac{PAR}{k\,PAR_{sat}}} \qquad\qquad (7)$$

F(680) and $F_{raw}(680)$ are corrected and uncorrected, respectively, PAR_{sat} corresponds to the light level capable to induce fluorescence saturation (evaluated by taking into account regional average PAR values) and k is a constant. The effectiveness of this correction is clearly shown in unvarying off-shore seawaters where the phytoplankton fluorescence emission is strongly affected by PAR during daily cycles (Figure 8), decreasing during higher irradiance values and increasing during nighttime.

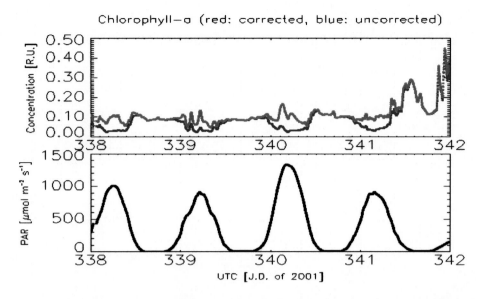

Figure 8. Comparison between continuous chlorophyll-a fluorescence emission and PAR (Indian Ocean, Dec. 4-7, 2001).

HYDROGRAPHIC LIF APPLICATIONS

Monitoring the marine ecosystem is a problem of primary importance in environmental control. In particular, coastal waters are considered major targets for this kind of surveillance, due to the possible presence of industrial wastes, including dangerous organic pollutants (PCB, dioxins, PAH) or crude oil discharges/leakages and anthropogenic releases (DOM and detergents). Such additional substances can heavily alter the coastal ecosystem by causing pollution and eutrophication situations, both factors affecting the growth of phytoplankton species which are at the origin of the marine food chain.

In this connection, lidar fluorosensors are emerging as very reliable instruments to remotely collect information about fluorescent targets for sea diagnostics purposes.

In the following, we will summarize some of the more significant examples in ecosystem monitoring LIF applications. Such optical tools can be also profitably employed in listening to early warnings of imminent ecological catastrophes and for prompt detection of changes in aquatic ecosystems, in real time and over large areas.

Oceanographic Monitoring

The Earth's oceans play a major role in climate equilibrium. Photosynthesis in aquatic systems is considered to be responsible for more than 40% of the global carbon fixation on an annual basis, by converting light radiation into organic compounds. At present, considerable uncertainties still exist in understanding the processes that control artificial and natural CO_2 uptake.

In particular, the key question concerning whether the Southern Ocean is able, like the North Atlantic Ocean, to take up atmospheric carbon dioxide, is still open to scientific discussions.

As already described, the ELF has joined different Italian expeditions in Antarctica. In particular, the XVI oceanographic campaign, 2000-2001 austral summer, supplied a unique opportunity to compare, for the first time, LIF Chl-a and pCO_2 surface distributions simultaneously, operating in this harsh environment [38]. In Figure 9, the anticorrelation between spot pCO_2 determinations and the continuous stream of LIF data is clearly evident (R = -0.8). The LIF survey revealed the presence of high productivity in polynya regions and close to the ice shelf edges, showing the high variability of the Antarctic Ross Sea.

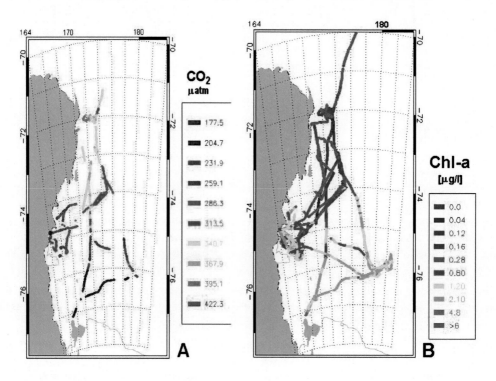

Figure 9. Maps of the distributions measured during the XVI Antarctic expedition; color scale indicates: a) pCO_2 [µatm] and b) Chl-a concentration [µg/l], average values are given for each color.

A summary of data measured at most significant sites crossed during the campaign is reported in Table 3, as collected by the Lidar apparatus (Chl-a concentration, and Y_{pp}), compared with both superficial pCO_2 values and the corresponding sea water mass. The data are presented in a chronological sequence.

The overall good agreement between two different experimental approaches confirms the reliability of the lidar fluorosensor apparatus as a *sea-truth* instrument for large area and real time analysis. Following the success of the present analysis, all data collected were jointly used as input to tune bio-optical algorithms for improving the precision of estimations of primary productivity (PP) in the Southern Ocean. A discussion of the application of standard models in Antarctica and more details on the model implemented can be found elsewhehere an example of the results of LIF based and standard PP models is given in Figure 10, based on the ELF measurements carried out during the 16th Italian expedition in Antarctica (January 5th 2001 – February 26th 2001). The present comparison indicates that usual PP models applied to standard chl-a concentrations can underestimate PP up to 50%.

Table 3 - Representative areas monitored during XVI campaign

Site	Lat-Lon	Date	Chl-a [$\Box$g/l]	pCO$_2$ [µatm]	Ypp
Southern Ocean	175°31.85'E 62°10.80'S	9/1/2001	0.11	373	0.40
TNB	164°11.59'E 75°51.70'S	14/1/2001	2.13	233	0.60
Mooring B	73°59.94'S 175°46.06'E	19/1/2001	0.20	319	0.28
Cape Adare	171°59.86'E 72°43.92'S	27/1/2001	0.15	449	0.11
Ross Isle	167°01.06'E 77°05.22'S	30/1/2001	2.45	177	0.08

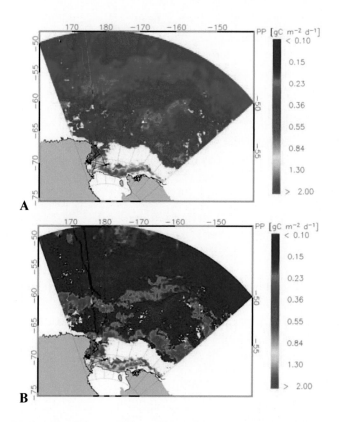

Figure 10. Average PP, based on the monthly products of January and February 2001, calculated with a) the ELF-calibrated SeaWiFS chl-a bio-optical algorithm and the new PP model, and b) the standard SeaWiFS chl-a bio-optical algorithm and the VPGM model by Behrenfeld and Falkowski. Continuous line: ship track.

Oil Slick Releases

The Venice lagoon and the nearby open sea area form an extremely peculiar ecosystem with high a risk of both industrial and anthropogenic pollution. Industrial activities carried out at Porto Marghera directly affect the composition of waters and bottom sediments with their liquid exhausts and gaseous emissions, while anthropogenic releases are expected to be of importance in the most populated areas of the lagoon (Venice center and Chioggia). The lagoon water flow system, characterized by limited exchanges with the open sea, makes difficult to eliminate industrial wastes together with all the anthropogenic organic substances and unburned oils used for combustion and local transport. Some of these substances, such as chlorinated poly-aromatic compounds, are characterized by a high persistence in water and a low probability of natural degradation under environmental agents (sun radiance, salinity and temperature). Due to the complexity of biochemical processes involved in the lagoon equilibrium, it is extremely important to monitor the local situation, with regard to the presence of specific pollutants which might generate high-risk situations.

Lidar systems may be used to monitor crude oils and components: their use is also common in identifying contaminated areas and to quantitatively determine the oil thickness dispersed on the sea surface. The spectral profile intensity of most common crude oils is a strong and broad fluorescence emission peaked at blue wavelengths, upon UV laser excitation [39]. In the case of LIF apparatus, operated at 355 nm in a large area, surface oil slicks can be identified from the water Raman scattering reduction, due to several contributions, including the oil and CDOM overlapping and radiation absorption by oil film on the water surface. Therefore, a precise underlying fluorescence and background measurement at the water Raman scattering emission has to be performed. The film thickness d is obtained rationing lidar Raman signals from the polluted area (R_{in}) with respect to clean open sea waters (R_{out}) [40] as:

$$d = \frac{1}{k_e + k_R} \ln \left(R_{in} \middle/ R_{out} \right) \tag{8}$$

where k_e and k_R are the oil and water Raman extinction coefficients, both in the case of assignment to CDOM and crude oil.

As an example, a large oil spot (Figure 11) was monitored inside an inner channel of the Venice lagoon, during the monitoring campaign performed by ELF systems installed on board of a small boat (10 m length, 3 m width and 1 m depth) in the overall lagoon and the surrounding sea-side.

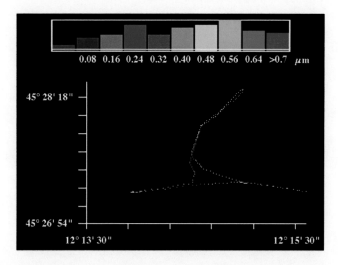

0.08 0.16 0.24 0.32 0.40 0.48 0.56 0.64 >0.7 μm

Figure 11. Oil slick measured in an inner Marghera channel of the Venice Lagoon (Nov. 1995).

Submarine Payload

As already reported in the introduction, surveillance and safeguard protection of marine environment concerns also the accidental or illegal discharges of polluted containers or drums, or releases from oil pipe lines. In this respect, LIF instruments again can play a key role in monitoring risk areas such as coastal waters or harbors, and ad-hoc submersible sensors can be installed for prompt detection of pollutant releases.

Moreover, a submersible lidar fluorosensor, operated in the range resolved mode, can remotely retrieve the suspicious matter and obtain its dispersion.

Therefore, the know-how gained by the ENEA team, in operating the ELF system in Antarctica, was applied to develop a lidar fluorosensor payload for submarine investigation (ROV) and to design a new lightweight flying payload for large surface monitoring to be installed on a remotely piloted vehicle (RPV).

The submarine lidar fluorosensor payload was designed for range resolved measurements, aimed to monitor vertical concentration profiles of dissolved organic substances and phytoplankton along water columns of several tens of meters. The apparatus was installed as an interchangeable package on the Antarctic ROV, built and tested at the Robotics Department of IAN-CNR for marine science applications [41]. The frame of the lidar payload consists of a stainless steel cage, supporting two titanium cylinders, each with a 300 mm diameter and 1100 mm length. The optoelectronic components, including a compact Nd:YAG laser operated in 3rd harmonic (Figure 12), and the receiver, have been redesigned in order to fulfill the ROV logistic requests and to operate in different environmental conditions, either in Antarctic seas or in the Mediterranean Sea.

Figure 12. Custom-designed Nd:YAG laser source.

The detailed characteristics of the apparatus are listed in Table 4, while Figure 13 shows the submarine payload with its external frame for connecting with its carrier. On the right side, the modular custom-designed Nd:YAG laser is installed.

Table 4 - Main characteristics of the submersible lidar fluorosensor

Transmitter	Nd:YAG laser
Pump	Energy=30 mJ
Probe	Energy=3 mJ
	Pulse length=10 ns
	Ppr 10 Hz
Expander	Variable (3x)
Detectors	Hamamatsu PMT R-1924 (2), R-1925 (2)
Filters	Dichroic T>90% (@ 400 nm)
	Interferential 402, 450, 650, 680 nm
Telescope	Cassegrain 23 cm dia. F#2
Fiber Optic	Plastic Multifiber = Four Branches
	Bundle Diameter Input 24.5 mm, Output 7 mm
	Length = 50 cm
Digitizer	Signatec ISA/PCI 500 Ms/s, 8 bit
Computer	Axiom AX6050DWP Passive Backplane

An internal aluminum cage hosts the laser and the high voltage power supply, to allow them to be easily extractable for external service. The estimated power consumption when the laser is running at its maximum power is in the order of 1 kW; therefore, two heat exchangers have been included for temperature control operation. The first one connected to the laser head, positioned downward the optical plane, should guarantee a proper operation in the Antarctic sea, while the second one should be used for the same purpose in tropical areas. In order to prevent freezing and damage to the laser components during hardware transfers, the internal water cooling system was evacuated through the input/exhaust connection of the first titanium cylinder. Internal sensors have been installed for monitoring the laser emitted energy, water temperature and humidity. An inert gas fluxing inside the tubes at high pressure prevents the formations of external arc. All the sensor signals, including the high voltage power supply, Q-switch and laser controls were transmitted to the computer through an RS232 interface cable. The second tube on the left hosts the send/receive optics (beam expander and telescope), the PMT detectors and the computer. The laser beam, passing through the small connection tube, is expanded and directed on the chosen optical path (horizontally, downwards or upwards) by means of an external assembly containing a large mirror. The LIF signals, after being filtered, were detected by the PMTs at the selected wavelengths and stored in the internal memory of four fast transient digitizer PC cards. An industrial PC hosting PCI/ISA slot cards was used to control all the experimental settings, as well as the acquisition and temporary data storage. Optical signals were transmitted and received through a large quartz window, allowed by a suitable rotation of the carrier, either to trace the CDOM at the marine bottom or to monitor the phytoplanktonic community below the pack ice. Electric power was supplied to the lidar payload by an external generator through an underwater umbilical cable, carrying also the data communication according to TCP/IP protocol specifications. The system overall weight of about 150 kg corresponds to the maximum allowed for a scientific payload to be carried with the used ROV.

The lidar fluorosensor payload was assembled at the Italian Antarctic base TNB (4/12/2001 - 16/01/2002), and tested inside a large tent placed over the icepack through an aperture 2m long, 1 m wide and 3.5 m thick. The tent also hosted all the electronics and mechanics for supporting the diving of ROV and its payloads (Figure 13).

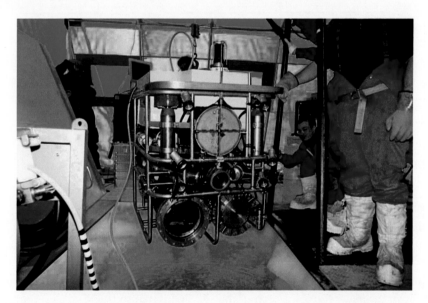

Figure 13. Lidar fluorosensor payload attached to the ROV, inside the tent (TNB, 24/12/2001).

Several complete vertical profiles were gathered up to the maximum reachable depth of approximately 200m, as shown in Figure 14, as obtained by integrating at the full range, at each underwater quote. The range integrated responses at 404nm (Raman backscattering) and at 450nm (CDOM) are reported in Figure 14A, while their ratio (CDOM normalized to

Raman signal) is shown in Figure 14B. Minor variations related to water characteristics are detected in all the curves down to about 100m, corresponding approximately to the lower end of the euphotic zone. At this depth a strong increase emerges in crossing a boundary layer, suggesting the presence of a vertical stratification between the upper and lower seawater levels. The observed variation could actually be related to changes either in the composition of optically active molecules, in CDOM total concentration or in other particle gradient, while all these phenomena may occur all together as well.

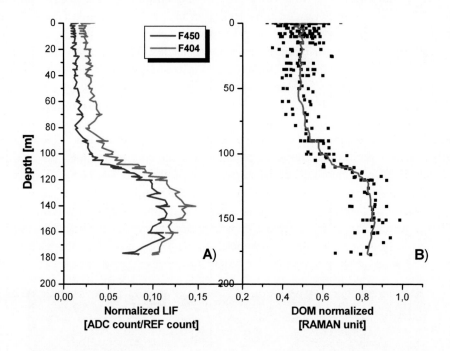

Figure 14. Vertical transect. Behavior of the integrated LIF signals along the depth: A) 404 nm and 450 nm; B) 450 nm signal Raman normalized [TNB, 13/01/2002].

CONCLUSION

The LIF technique is a powerful tool for qualitative and quantitative determinations. At the present time, its capabilities have not been fully exploited and with the development of new and miniaturized laser sources, such as the high power diode lasers, new and interesting applications will be investigated.

In the case of marine applications, the large database of fluorescence determinations, collected along the ship cruise, has been demonstrated to be fruitfully applied for successive calibration of satellite radiometer images. The final data merging strongly improves the estimation of chlorophyll-a and primary productivity of the oceanographic area investigated, and therefore allows gathering precise information on global changes.

REFERENCES

[1] http://en.wikipedia.org/wiki/Laser-induced_fluorescence

[2] de Mello, A.J. (2003). *Lab Chip*, 3, 29N.

[3] Castaldi, M.J. & Senkan, S.M. (1998). Real Time, Ultrasensitive Monitoring of Air Toxics by Laser Photoionization Time-of-Flight Mass Spectrometry *Journal of the Air & Waste Management Association*, 48, 77-81.

[4] Lichtenthaler H.K. & Rinderle U. (1988). The Role of Chlorophyll Fluorescence in the Detection of Stress Conditions in Plants. *CRC Crit. Rev. Anal. Chem.*, 19, supl. 1, S29-S85.

[5] Hoge F.E. & Swift R.N. (1981). Application of the NASA Airborne Oceanographic Lidar to the Mapping of Chlorophyll and Other Pigments. NASA Conf. Pub. "Chesapeake Bay Plume Study, 349.

[6] Diebel D, Hengstermann, T., Reuter, R. et al. (1989). Laser fluorosensing of Mineral Oil Spills. Remote Sensing of Maritime Pollution. AE Lodge (editor), J Wiley & Sons (Chichester), 165 pp, 127-142.

[7] Dai, Z. Tseng, L.K. & Faeth, G.M. (1995b). Velocity/mixture fraction statistics of round, self-preserving, buoyant turbulent plumes. *Journal of Heat Transfer-Transactions of the ASME.* 117:918–26.

[8] Neij, H., Johansson, B. & Aldén, M. (1994). Development and demonstration of 2D-LIF for studies of mixture preparation in SI engines. *Combust. Flame* 99:449-457.

[9] Sanders, S.J., Boivin, R.F., Bellan, P.M., et al. (1999). Added discussion of "Observations of fast anisotropic ion heating, ion cooling, and ion recycling in large-amplitude drift waves", *Phys. Plasmas.* 6, 4118.

[10] Maslov, D.V. Fadeev, V.V. & Lyashenko A.I. A Shore-Based Lidar for Coastal Seawater Monitoring. EARSeL-SIG-Workshop LIDAR, pag. 46-52.

[11] Cecchi, G., Pantani, L., Breschi, B., et al. (1992) "FLIDAR: a Multipurpose Fluorosensor-Spectrometer" EARSeL *Advances in Remote Sensing* 1, 72-78.

[12] Johansson, J., Wallinder, E., Edner, H., et al. (1992). *Fluorescence Lidar Multi-Color Imaging of Vegetation*, Proc. 16th ILRC, NASA CP 3158, 433-436.

[13] Reuter R, Wang, H., Willkomm, R., et al. (1995). A laser fluorosensor for maritime surveillance: Measurement of oil spills. EARSeL *Advances in Remote Sensing*, 3, 152-169.

[14] Hoge, F.E. (1983). Oil film thickness using air-borne laser-induced fluorescence back-scatter. Appl. Opt. 22: 33.3318.

[15] Babichenko, S., Kaitala, S., Leeben, A., et al. (1999). Phytoplankton pigments and dissolved organic matter distribution in the Gulf of Riga. *Journal of Marine Systems* 23(1-3): 69-82.

[16] Reuter, R., Willkomm, R., Krause, G., et al. (1995). Development of a Shipboard lidar: Technical layout and first results. EARSeL *Advances in Remote Sensing,* 3, 15-25.

[17] Barbini, R., Fantoni, R., Colao, F., et al. (1999). Shipborne laser remote sensing of the Venice lagoon. *Int. Jou. Remote Sensing*, Vol. 20, N° 12, 2405-2421.

[18] Drozdowska, V., Babichenko, S. & Lisin, A. (2002). Natural water fluorescence characteristics based on lidar investigations of a surface water layer polluted by an oil film; the Baltic cruise – May 2000 OCEANOLOGIA, 44 (3), pp. 339–354.

[19] Harsdorf, S, Janssen, M., Reuter, R., et al. (1999) Submarine lidar for seafloor inspection. *Measurement Science and Technology*, 10, 1178 – 1184.

[20] Barbini, R., Colao, F., Fantoni, R., et al. (2002). Submarine and Airborne lidar fluorosensor payloads for the remote sensing of the Antarctic Ross sea. Proceedings of the IATICE (Italian Australian Technological Innovations Conference & Exhibition) Melbourne 25-28 March 2002, The Embassy of Italy in Camberra, Australia, pag. 219-222.

[21] Measures, R.M. (1984). Laser remote sensing.Wiley-Interscience Publications, New York.

[22] Slusher, R.B. & Derr, V.E. (1975). "Temperature dependence and cross sections of some Stokes and anti-Stokes Raman lines in ice Ih," Appl. Opt. 14, 2116–2120.

[23] Bristow, M., Nielsen, D. Bundy, D., et al. (1981). "Use of water Raman emission to correct airborne laser fluorosensor data for effects of water optical attenuation," Appl. Opt. 20, 2889–2906.

[24] Barbini, R., Colao, F., Fantoni, R., et al. (2001). Differential lidar fluorosensor system used for phytoplankton bloom and seawater quality monitoring in Antarctica, *International Journal of Remote Sensing* 22, 369-384.

[25] Barbini, R., Colao, F., De Dominicis, L., et al. (2004). Analysis of simultaneous chlorophyll measurements by lidar fluorosensor, MODIS and SeaWiFS, *International Journal of Remote Sensing* 25, 2095-2110.

[26] Barbini, R., Colao, F., Fantoni, R., et al. (2001). Remote sensing of the Southern Ocean: techniques and results, *Journal of Optoelectronics and Advanced Materials* 3, 817-830.

[27] Barbini R., Colao F., Fantoni R., et al. (2003). Lidar fluorosensor calibration of the SeaWiFS chlorophyll algorithm in the Ross Sea, *International Journal of Remote Sensing* 24, 3205-3218.

[28] Barbini R., Colao F., Fantoni R., et al. (2003). Remotely sensed primary production in the western Ross Sea: results of in situ tuned models *Antarctic Science* 15, 77-84.

[29] Barbini R., Colao F., Fantoni R., et al. (2005). Lidar calibrated satellite sensed primary production in the Southern Ocean, *Journal of Optoelectronics and Advanced Materials* 7, 1091-1102.

[30] Falkowski, P.G., & Raven, J.A. (1997). *Aquatic photosynthesis*. Blackwell Science ISBN 0-86542-387-3.

[31] Cerullo, M.M. & Curtsinger, B. (1999). *Sea Soup: Phytoplankton*. Tilbury House Publishers ISBN: 0884482081.

[32] Kirk, J.T.O. (1994). Light and Photosynthesis in Aquatic Ecosystems. Cambridge: Cambridge University Press, ISBN 0-521-45966-4.

[33] Sournia (1979). Phytoplankton Manual (Monographs on oceanographic methodology) Unipub eds. ISBN: 9231015729.

[34] Raymont (1980). Plankton and Productivity in the Ocean, Pergamon International Library of Science, *Technology*, ISBN: 0080215521

[35] Dring, M.J. (1982). *Biology of Marine Plants*. Edward Arnold Publisher, ISBN: 0713128607.

[36] Chekalyuk, A.M., Demidov, A.A., Fadev, V.V., et al. (1992). Laser remote sensing of Phytoplankton and organic matter in the sea water. In International Archives of Photogrammetry and Remote Sensing, L. W. Fritz, J. R. Lucas, Eds., Proceedings XVII *International Congress for Photogrammetry and Remote Sensing*, 29: 878-885.

[37] Barbini, R., Colao, F., Fantoni, R., et al. (1997c). Photosynthetic activity and electron transport measurements using laser pump and probe technique. *Remote Sensing Reviews*, 15: 323-342.

[38] Barbini, R., Ceradini, S., Colao, F., et al. (2003). Simultaneous measurements of remote lidar chlorophyll and surface CO2 distribution in the Ross Sea *Int. J. Remote Sensing*, vol. 24, n° 1, 1-13.

[39] Fantoni, R., Barbini, R., Colao, F., et al. (1994). Applications of excimer laser based remote sensing systems to problems related to water pollution. In Excimer Lasers, L.D. Laude Ed., NATO ASI series E-vol. 265, (Dordrecht: Kluwer Academic Publ.), pp 289-305.

[40] Hoge, F.E. & Kincaid, J.S. (1980). Laser measurement of extinction coefficients of highly absorbing liquids, *Applied Optics*, 19, 1143-1150.

[41] Veruggio G., Caccia M. & Bono R. (1997). "ROMEO: a multi-role unmanned underwater vehicle for advanced robotics developments", 4th International Symposium on MMAR97 Methods and Models in Automation and Robotics, Miedzyzdroje, Poland, Agosto, vol. 2, pp. 517-522.

3. LASER APPLICATION IN SOIL MONITORING

In: Laser Applications in Environmental Monitoring
Editors: L. Fiorani and F. Colao

ISBN 978-1-60456-249-1
© 2008 Nova Science Publishers, Inc.

Chapter 3.1

SOIL ANALYSIS BY LASER INDUCED BREAKDOWN SPECTROSCOPY

Violeta Lazic

Laser Applications Section, Advanced Physical Technologies and New Materials
Department, Italian National Agency for New Technologies, Energy
and the Environment (ENEA),
via Enrico Fermi 45, 00044 Frascati, Italy
E-mail: violeta.lazic@frascati.enea.it, tel.: +39 – 06-94 00 58 85

ABSTRACT

Laser Induced Breakdown Spectroscopy (LIBS) is a relatively recent, non-contact technique for rapid multi-elemental analysis of solid, liquid and gaseous samples and in different ambient conditions. It is also widely used for soil analysis, including monitoring of the elements important for the environmental protections. In this chapter, the reader is introduced to the LIBS technique, to its advantages and limitations, and to the instruments with their principal components and typical configurations. The main characteristics of the laser-produced plasma are discussed as well as a way to obtain the sample information from the spectrally resolved plasma emission. Qualitative sample characterization is initially described, and then the review continues by considering more complex quantitative analysis. Different factors affecting the measuring accuracy are discussed and the methods for minimizing the influence of some experimentally uncontrollable variables. Common approaches adopted for retrieval of the element concentrations in the sample are described and critically analyzed. To this aim several examples are presented, some of them not regarding the soils although explanatory for the LIBS technique in general.

In a separate section, the attention is devoted to soil and sediment analyses where various topics of the LIBS application are discussed, including the experimental conditions, the analyzing method, the final results and data interpretation. These examples also include the soil pollutants analysis, representing however only a special case of the LIBS elemental analysis. Application to planetary exploration is also reviewed since it concerns soil/rock analysis and gave results of some importance also for the characterization of these materials in Earth conditions. In this context, the influence of the surface temperature on the LIBS signal behavior is discussed, reporting the results obtained in the simulated Martian conditions but the similar effects are also expected for the measurements in the atmospheric air.

1. PRINCIPLES OF LIBS

Laser Induced Breakdown Spectroscopy (LIBS) technique [Cremers et al. 2006] is a powerful tool for in-situ elemental measurements in different surroundings such as vacuum, then gas and liquids at different ambient pressures. The technique is based on plasma generation by an intense laser pulse, which duration is in nanosecond range or shorter. When analyzing the gasses or bulk liquids, the laser pulse generates a breakdown in the media. In the case of solid samples, the plasma is produced through laser-induced evaporation of the surface layer [Piepmeier 1986]. In both the cases, the intense laser pulse is also responsible for atomization and ionization of the material.

The plasma growth and decay lead to different processes such as: expansion; shock waves formation; continuum (Bremsstrahlung) emission and light absorption by free electrons (inverse Bremmstrahlung); collisions in the gas cloud with excitation and relaxation of atoms/ions; chemical recombination and, as important for LIBS, de-excitation of the species (atoms, ions and molecules) through emission of the radiation. Detection of the latter emission, obtained after laser-induced plasma formation, is a principle of LIBS technique. This radiation is usually detected in the spectral range covering near UV, visible and near IR.

Initially, the plasma temperature is very high, typically above 15000 K, and its radiation is dominated by the continuum component (Figure 1.1) due to Bremmstrahlung emission from free electrons and ion-electron recombination [Griem 1964]. The continuum emission is higher for the measurements in more dense medium (high gas pressure or inside liquids), and has a relatively fast decay. It is followed by appearance of the ionic and atomic lines, where the ionic lines are more intense in the early plasma stage characterized with a higher temperature. Due to the high electron density in the early plasma, lines emission are strongly collisionaly broadened thus not allowing to resolve emission lines from most elements in the plasma. Often, the progressively decaying plasma emission is still detected after few tens of microseconds from the laser pulse. In sufficiently cooled plasma, the emission from the molecules, formed from the species initially present in the plasma, can be also detected. However, some molecules or fragments such as CN can be observed also for the shorter acquisition delays. In order to avoid that the initial strong continuum masks emission lines, and/or that the strong initially broadened lines cover the weaker lines, it is preferable to delay the spectral acquisition with respect to the laser pulse. The acquisition gate and delay in certain experimental conditions can be optimized for obtaining maximum Signal-to-Noise Ratio (SNR) for a range of the analytical lines [Wisbrun et al. 1994, Colao et al. 2004].

The atomic and ionic lines, once assigned to specific transitions given in databases [NIST], allow for a qualitative identification of the species present in plasma and consequently of the elements initially present in the sample. The relative intensities of the emission lines can be used for the quantitative determination of the corresponding elements (see section 6).

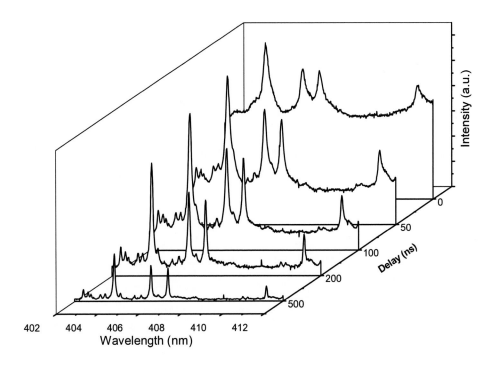

Figure 1.1. Example of LIBS spectra from stainless steel as a function of acquisition delay from the laser pulse. The signal acquisition gate is 50 ns.

A growing interest in LIBS technique and its rapid development in the past two decades can be attributed to its numerous advantages such as:

- Possibility to perform in-situ measurements with minimum or none sample preparation, and in different surroundings
- Theoretically, all the elements could be contemporary analyzed, both in traces and for the concentrations up to 100%. Development the of a wide range, high-resolution spectrometers (see section 2) allowed to capture the whole spectra necessary for the sample characterization, which often can be obtained by applying even a single laser shot
- The technique is essentially non destructive i.e. micro-destructive: as for example, ablation of less than one microgram of the sample could be sufficient for the analyzes.
- The measurements can be applied both for a single point analyses and for the extended surface mapping
- Through the material removal rate (typically less than 1 μm per pulse), it is possible to measure a vertical element distributions in a small scale. Laser ablation can be used also for removal of the unwanted surface layers before than the analyzes.

- The instrument needs only an optical access to the sample, making possible to implement the measurements also in hostile environments. The optical head can even be installed also into a driller for on-line analyses during the soil perforations.
- Data analyses can be completely automatic and the instrument can be implemented for on-line analyses
- The cost of the instrumentation is relatively low and it is possible to achieve the system miniaturization
- Using an UV laser excitation, the system could be easily switched from LIBS to LIF (Laser Induced Fluorescence) measurements; the latter are important for biological studies

A wide range of LIBS applications has been developed in recent period [Cremers et al. 2006, Winefordner et al. 2004, Radziemski 2002], including: monitoring of environmental contaminations (soils, water, aerosols and biological tissues), control of material processing (particularly developed for steel industry), sorting of materials, monitoring of traditional and nuclear power plants, detection of explosives [De Lucia et al. 2003] and other hazardous materials, characterization of archeological materials and artworks [Giakoumaki et al. 2006] etc. Recently, LIBS instruments have also been proposed for planetary exploration (see section 6.5).

2. INSTRUMENTATION

A typical LIBS apparatus is shown schematically in Figure 2.1, and its main components are:

- Pulsed laser that generates plasma
- Focusing optics for the laser beam
- Sample holder (if used)
- Optical system for collecting the plasma emission and for transporting it to a spectrometer. It can contain lenses, mirrors and fiber optics.
- Spectrometer for light dispersion, equipped with a detector – usually CCD device is used without or with an intensifier (ICCD).
- Computer and electronics for triggering the laser, synchronization with the detector, eventual gating of the detector (for ICCD), then for the data acquisition and storage.

2.1. Lasers

For LIBS it is necessary to use the laser pulses with a high peak power, typically of 5 MW or more. Such a peak power can be obtained with moderate laser pulse energies (in order of 1 mJ) which have duration in nanosecond range. More efficient plasma generation is obtained with shorter pulses, as for example of femtosecond duration. However, the complexity and costs of femtosecond lasers make them less attractive for in-field applications.

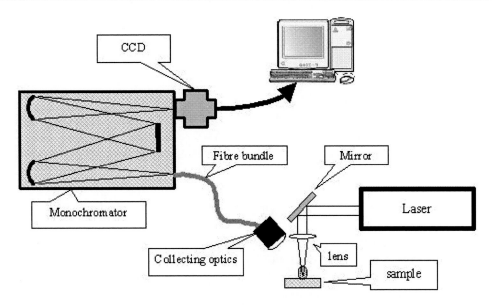

Figure 2.1. A common LIBS lay-out: bi-axial configuration.

Monochromacity of the laser source for LIBS is not critical. Regarding the laser wavelength choice, the following considerations must be taken into account:

- Sample absorption at a given laser wavelength: for example, the metals are generally more reflective at longer wavelengths; consequently their ablation requires higher laser peak powers. It is indeed a general rule that most of the materials have stronger absorption at shorter laser wavelengths.
- The surrounding medium should be transparent for the laser radiation: for underwater analyses it is preferable to use green wavelengths where water has a transmission window
- For shorter laser wavelengths, the continuum component of the LIBS plasma is lower due to reduced (inverse Bremmstrahlung) absorption of the laser radiation by free electrons in plasma. Lower continuum emission is favorable when using a compact and cheaper CCD detector without gating option.

A laser beam with high beam quality and short wavelength can be focused onto a smaller spot. This is especially important if a micro-sampling is required, and to achieve high energy density in the focal point. The latter can be critical when using the laser sources with very low pulse energy.

Beam quality of the laser is important for obtaining small spot sizes, and it is particularly critical for long-range (stand-off) measurements where the beam is focused at distances up to 100 m [López-Moreno et al. 2004]. In the latter case, also the beam directionality becomes very important.

Pulse-to-pulse laser stability is not critical for the signal accumulation over a sufficient number of laser shots. It becomes more critical in the case of single shot quantitative analyses without applying some signal normalization (see section 6).

In all the cases, the choice of the laser source is generally guided by its complexity, requirements for the maintenance, dimensions and weight, cooling and power requirements, rugged design if used for in-field operations, and finally by the costs.

In the majority of LIBS measurements a flash-lamp pumped, Q-Switched Nd:YAG laser is used. This type of laser has a widespread diffusion also for industrial applications because of its numerous advantages over the other laser sources. Among them, there is also the possibility to operate at different wavelengths (1064 nm, 532 nm, 355 nm and 266 nm), which is of interest for research. Often, all these wavelengths could be changed inside the single laser source, by an appropriate insertion of the frequency conversion elements. Beside, two or more laser pulses can be extracted during a single lamp flashing [Lazic et al. 2005, Balzer et al. 2005], with advantages for the analytical LIBS performance [Babushok et al. 2006].

Researches on the LIBS system miniaturization and with reduced power consumption have brought to the implementation of diode pumped lasers containing a passive Q-Switch [Gornushkin et al. 2004, Zayhowski, 2000]. In absence of the amplification stage, these lasers can produce the nanosecond pulses with energy up to 2 mJ. Due to low pulse energy, a tighter focusing is required in order to achieve sufficient SNR.

2.2. Optical System

Laser beam, sometimes expanded, is focused onto a sample by means of one or more focusing lenses. In some cases, instead of lenses a telescopic system containing the mirrors is employed [López-Moreno et al. 2004]. Focal length of the system must provide a sufficiently small laser spot, i.e. the sufficient laser energy density for the plasma generation. Using low energy lasers, such as microchip lasers [Gornushkin et al. 2004], it could be necessary to employ a microscopic objective for the beam focusing. For underwater LIBS applications, a lens (or lenses system) with a short focal length must be used (typically 20-30 mm) due to the high water absorption and light scattering by the eventual suspended particles [Lazic et al. 2007].

Beside, a bi-axial configuration for the laser beam focusing and plasma light collection (Figure 2.1), a mono-axial system is often used (Figure 2.2). In such a case, the beam usually passes a pierced mirror, which serves to deflect the plasma radiation towards the detection system. Alternatively, a dielectric mirror for the laser beam deflection can be used, and the plasma emission is then collected through the mirror. However, this type of mirror can cut the signal in certain spectral intervals important for the analysis of some elements. The mono-axial configuration is less sensible to misalignment and particularly recommended for analyses of irregular surfaces such as rocks and soils in natural state.

For most of the sample types, the elemental emission lines in UV spectra supply a lot of analytical information, so the collecting optical system should be transparent in this spectral range. In these cases quartz optics is usually employed.

Typical collecting system contains an optical fiber or a fiber bundle. In the latter case, the exit of the bundle is arranged into a line, to match or to replace the monochromator entrance slit. On the other hand, compact spectrometers (see the next section) and Echelle spectrometers generally have an entrance for a circular fiber. In particular, the Echelle systems have useful aperture only up to 50 μm diameter. If an array of the compact spectrometers (up to eight) is used, a fiber bundle serves to split the optical signal towards the entrances of different spectrometers.

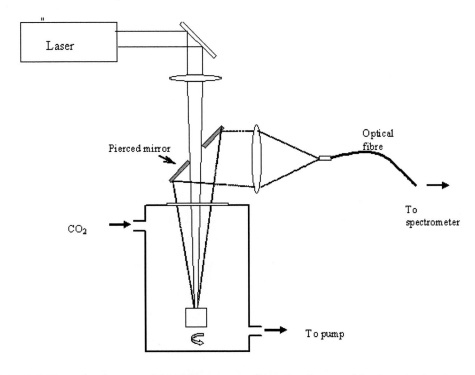

Figure 2.2. Example of mono-axial LIBS system configuration, here used for detection in a low pressure chamber.

2.3. Spectrometers and Detectors

As it was mentioned earlier, the plasma spectrum contains emissions from the continuum component, the atomic, ionic and eventually molecular species. The emission lines become narrower with increasing of the acquisition gate delay as the plasma expands and collisionaly broadening is consequently reduced. On some type of the materials, as for example most of the soils, the LIBS spectrum is very populated with the emission lines belonging to different elements. Many of these lines, necessary for the element identification in the plasma i.e. in the sample, suffer from the spectral interference. As for example, in the spectral range from 200 nm to 800 nm, typically considered for multi-elemental LIBS analyses, accordingly to the NIST data base [NIST], there are 2266 lines of the atomic iron and 1012 from its first ionization stage. Although many of these lines are very weak and below the detection threshold, it is clear that iron lines in the samples reach of this element, can mask or interfere with the lines from many other elements. Similarly, many other elements exhibit a rich spectrum, and the number of the detected lines depends on the corresponding element concentration in the sample. Due to spectral interferences, it is important to use high-resolution spectrometers, a 0.1 nm or better resolution is typical for the LIBS systems. In Figure 2.3 there is and example of LIBS spectra obtained by a low resolution and a high resolution system, in the latter case with a 0.55 m monochromator equipped with a grating

1200 gr/mm. The loose of the analytical information is clear in the case of a low resolution system, which generally allows for the detection of the main sample constituents but rarely for minor and trace elements. This fact is not only due to the spectral overlap of different lines, but also due to reduced SNR (see Figure 2.3), critical for the weak emission lines such as expected for the elements present in the low concentrations.

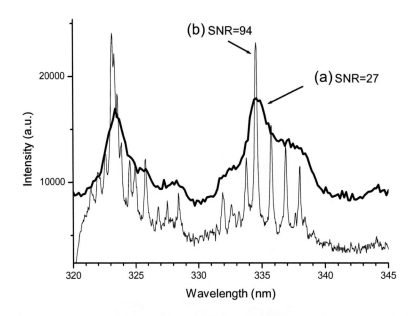

Figure 2.3. A single-shot LIBS spectrum from a sediment recorded: (a) by a Low-Resolution system (0.3 nm); (b) by a High-Resolution system (0.07).

New designs of the high resolution spectrometers have been developed for LIBS, in particular the compact spectrometers [Stellatnet] and Echelle [Hamilton et al. 2006]. Compact spectrometers (Figure 2.4), also existing in a low resolution version for other spectroscopic techniques (example for LIF), incorporate array detectors and up to eight modules are used to cover the spectral range between 200 nm and 1000 nm. These modules are synchronized between them and with the laser pulse. The detectors employed in the compact spectrometers (linear Charge Couple Device - CCD or Photo Diode Array - PDA) do not allow for controlling the gate width, and the integration time is of order of 1 ms. This fact is disadvantageous in several cases:

- In presence of an intense background illumination, as for example during the measurements with daily light: the long integrated continuum light partially masks the LIBS signal of a relatively short duration (in order of 1-10 μs).
- For detection of short living species, which recombine in the plasma
- For detection of the plasma in Local Thermal Equilibrium (LTE), often necessary for the plasma modeling (section 3).

- For underwater measurements, sonoluminescence emission [Lohse 2002] can occur following the collapse of a vapor bubble formed after the breakdown [Baghdassarian et al. 1999]. The bubble collapse appears later than the LIBS signal and a very strong and spectrally broad sonoluminescence sometimes completely masks LIBS signal [Lazic et al. 2007].

Conventional spectrometers often have a CCD (with or without an intensifier) for recording the diffracted light. The CCD can be used as an image detector for measuring the spectrally resolved light distribution along the slit. In LIBS analyzes the CCD is usually operated in full vertical binning. In this way, spatial light distribution along the slit is lost, but much higher signal is obtained due to vertical (parallel to the slit) signal integration across the pixels. Detectors typically have a larger number of pixels (example 1024) in direction normal to the slit so to cover a wider spectral range, while the smaller number of pixels (example 256) in the other direction is sufficient to capture the whole slit image at the exit spectrometer plane.

Figure. 2.4. Example of a compact, high-resolution spectrometer [Stellarnet].

Echelle spectrometer (Figure 2.5) employs square CCD (for example 1024x1024 pixels). This spectrometer contains cross-dispersive elements to produce highly resolved spectrum in one direction and then to separate different grating orders along the perpendicular axis [Hamilton et al. 2004, Florek et al. 2001]. In this way, the image recorded at the exit plane, contains various orders corresponding to different wavelength ranges. For spectral analyses, a specific software is required to transform so formed image to the spectral light distribution. With the commercial Echelle spectrograph, the full spectra can be recorded in the range of 200-800 nm or 200-1000 nm, and with a resolution of 0.05-0.1 nm. Contemporary recording of the full spectra drastically reduces the measuring time for multi-elemental analyses. As for example, when using a scanning monochromator, the spectral range from above requires at least 30 measurements. Simultaneous detection of the complete spectrum is necessary for a single sample point characterization, as for example: during the multi-elemental surface mapping or vertical sample profiling, then for characterization of small targets (example droplets) and for micro-destructive analysis as is the case of artistic and historical objects. Furthermore, single-shot full spectra detection allows to observe shot-to-shot signal variations, and to implement for different kinds of data reduction.

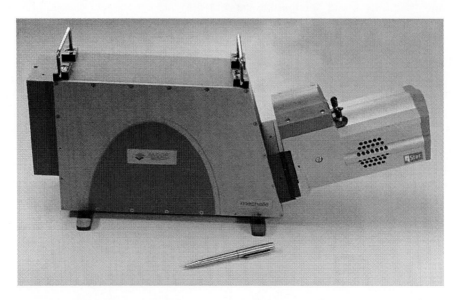

Figure 2.5. Echelle spectrometer [Andor].

However, Echelle spectrometer has some disadvantages, the most important beside a high cost, is a relatively low signal. Reduced signal with respect to the other spectrometers with a CCD is due to integration over only 3-5 pixels while in other cases a full vertical binning (example over 200 illuminated pixels) is generally applied. Such limited pixel integration is necessary to avoid an overlap with the orders diffracted in close region of the detector. Therefore, the spectrometer entrance slit must also be small – with diameter up to 0.05 mm. This leads to reduced signal collection efficiency as a large portion of the plasma is imaged out of the slit aperture, directly or through an optical fiber.

Commercial Echelle spectrometers contain a cross-dispersing prism (or prisms). Long duration measurement runs require the spectrometer thermal stabilization, since the small temperature variations can cause a shift between different diffraction orders. Otherwise, repeated wavelength calibrations would be required.

The most advanced and flexible LIBS detection system is based on an ICCD, which allows both for the measurements of the spatial and temporal light distribution spectrally resolved. The detector can be gated down to a few nanoseconds and is used for determining of the temporal plasma behavior. A main disadvantage of an ICCD is a relatively high cost and larger dimensions, so it is not used for the compact LIBS systems. Comparison of the CCD and ICCD based system for LIBS applications are reported in [Carranza et al. 2003].

For a limited number of analyzed emission lines, instead of spectrometers it is possible to use narrow band-pass filters coupled with other detectors such as a photomultiplier (PMT), photodiode (PD), avalanche PD (APD), etc. Such types of detection systems are generally implemented for some very specific applications.

3. PLASMA CHARACTERISTICS

LIBS diagnostics, and in particular quantitative LIBS analysis, are often based on some assumptions, the most common being that the detected plasma is in Local Thermodynamic Equilibrium (LTE), and sometimes that the plasma is optically thin. Starting from these assumptions, the plasma parameters (temperature and electron density) can be determined and used for further data processing.

3.1. Plasma in Equilibrium

If the plasma is in equilibrium, the ionization depends only on the plasma conditions (density and temperature) through Saha equation and the excitation is determined by Boltzmann statistics [Griem 1964]. Equilibrium occurs in high-density plasma where collisions dominate over radiative processes. Low-density plasmas have often a charge state, which is significantly lower than the equilibrium value.

The thermodynamic equilibrium is rarely complete, so the concept of LTE is introduced, which means that the equilibrium occurs in a limited plasma volume and it can be different from region to region. In the case of LIBS plasmas, often is assumed that observed plasma volume, centered to the hot plasma core, is in LTE. However, the external plasma regions are known to be cooler and less dense and part of their emission is also collected together with the emission coming from the central, more uniform plasma.

Let us assume that the plasma contains only electrons (e), atoms and their first ionization stages, as it is usually case in LIBS plasmas during the detection window. Both heavy particles and electrons have Maxwell distribution corresponding to their temperatures: T_h and T_e respectively, where T_h depends on the particle mass. If the plasma is in LTE, the electron temperature and the temperature of heavy particles are identical ($T_h = T_e \equiv T$), and the level population for all atoms/ions is given by Boltzmann formula, which includes the unique plasma temperature:

$$\frac{N_k}{N_i} = \frac{g_k}{g_i} e^{-(E_k - E_i)/kT} \tag{3.1}$$

Where k, i refer to two levels with the corresponding populations N_k and N_i, which energies are E_k and E_i respectively. The corresponding level degeneracy (statistical weight) is g_k, g_i and they are given in the atomic data-bases. Here, k is Boltzmann constant and $U(T)$ is partition function for a single species (tabulated or calculated values).

In case of optically thin plasma in LTE, the recorded spectrally integrated line intensity I^{ki} corresponding to the atomic or ionic transition between energy levels E_k and E_i of the species α (atomic or ionic) is given by:

$$I^{ki}_\alpha = a'_\alpha N_\alpha \frac{g_k A_{ki} e^{-E_k/kT}}{U_\alpha(T)} \tag{3.2}$$

Where a'_α is a constant depending on the experimental conditions, N_α is species number density in plasma, A_{ki} is the transition probability. If the LIBS instrument is radiometrically calibrated, the atomic data accurate and complete, and the whole plasma volume imaged into detector or the species have the same distribution in the plasma, the constant a'_α is the same for all the species and independent of the line wavelength.

For plasma in LTE the relative populations among ion stages are given by the Saha equation [Griem 1964]. In typical LIBS plasmas, during the observation window, the presence of the second and higher ionization stages could be neglected. Accordingly to the Saha equation, the number density ratio of neutral (N_α^I) and first ionized species (N_α^{II}) of each element α in the plasma depends on plasma temperature and electron density N_e:

$$\frac{N_\alpha^{II}}{N_\alpha^I} = \frac{1}{N_e} \cdot \frac{U_\alpha^{II}(T)}{U_\alpha^I(T)} B(kT)^{3/2} e^{-\frac{E_\infty}{kT}} \equiv f_{2\alpha}(N_e, T) \tag{3.3}$$

where: $U_\alpha^I(T)$ and $U_\alpha^{II}(T)$ are partition functions of atomic and the first ionization stage respectively; E_∞ is the effective ionization energy in the plasma surrounding; B is a constant with a value of 6.05E+21 cm^{-3}.

LTE requires that collisionaly excitation and de-excitation electronic rates dominate over radiative processes, a condition achieved if the plasma electron density is satisfying the relation [Huddlestone et al 1965]:

$$N_e(cm^{-3}) \geq 1.6 \cdot 10^{12} \cdot \sqrt{T(K)} \cdot \Delta E^3 \tag{3.4}$$

Where $T(K)$ is the plasma temperature and $\Delta E(eV)$ is the largest energy gap for which LTE conditions hold. From this expression, it is clear that LTE requires high electron densities (typically 10^{16} cm^{-3} or higher), a condition often achieved in LIBS plasma generated at atmospheric or higher ambient pressures.

3.2. Plasma Temperature

There are different methods for measuring the plasma temperature from its spectral emission [19]; however, the most used approach is based on the intensity ratio between two or more lines belonging to the same species. Starting from the Equation (3.2), the intensity ratio (measured) of two lines from the same species α (atomic or ionic) is:

$$\frac{I_{ki}'}{I_{ki}} = \frac{g_k' A_{ki}'}{g_k A_{ki}} e^{-(E_k' - E_k)/kT} \tag{3.5}$$

By measuring the integral line intensities obtained by the instrument previously corrected for the spectral response, and for the lines with known values of A, g and E, it is possible to calculate the plasma temperature. The chosen lines must not be self absorbed otherwise the calculated values could be quite erroneous.

The two lines method for the temperature determination is not precise, and the accuracy can be improved by including simultaneously more lines of the same species into the calculation. To evaluate the plasma temperature, we take the natural logarithm of equation (3.2) obtaining:

$$\ln\left(\frac{I^{ki}_\alpha}{g_k A_{ki}}\right) = -\frac{E_k}{kT} + \ln\left(\frac{N_\alpha a_\alpha}{U_\alpha(T)}\right) \tag{3.6}$$

By plotting the left hand term of equation (3.6) as a function of E_k, different emission lines emissions belonging to the same element lie along a straight line with a slope of $-1/kT$. This is called Boltzmann plot and for obtaining an acceptable accuracy in the temperature calculation, it is crucial that the spectroscopic constants appearing in eq. (3.6) are accurate, that the upper levels are sufficiently spaced and inside limits of LTE (see Equation 3.4) and that the considered lines are not saturated.

An example of Boltzmann plot is shown in Figure 3.1, obtained from a LIBS spectra of one copper alloy. The measurements were performed in air. The considered lines belong to atomic Cu, and in order to avoid very intense lines that could be self-absorbed (particularly the resonant transitions), only the lines with relative intensities I<3000 according to the NIST database [NIST] were included in the Boltzmann plot. The temperature calculation was performed automatically, after fitting the whole spectra (240-750 nm). The details are discussed in [Colao et al. 2002].

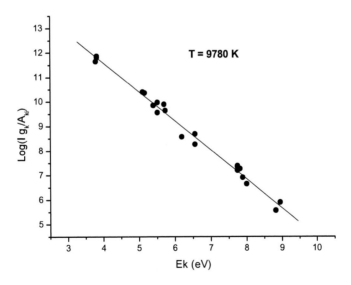

Figure 3.1. Example of the Boltzmann plot and the calculated plasma temperature.

Plasma temperature at early times is very high, and auto-ionization is dominant. The fast collisionaly processes bring the plasma to the conditions of LTE where the electrons have Maxwell distribution, all the species have the same temperature, the energy levels populations correspond to Boltzmann distribution, while the Saha equation describes the concentration ratio between two successive ionization stages. Further plasma cooling, expansion and recombination cause plasma departure from LTE. One of the indicators for missing LTE conditions is the Boltzmann plot. In Figure 3.2 there is an example of the Boltzmann plot measured at different delays from the laser pulse [Colao et al. 2002b]. A straight line, indicating the Boltzmann population, fits well the data point only for delays between 100 ns and 1500 ns. For other delays, there is a loss of LTE. As for example, for delay of 2000 ns, fitting the points corresponding to the energy levels below or above 5 eV, two different temperatures would be obtained. This indicates that the electron density is too low for LTE existence over the whole energy span (3-7 eV).

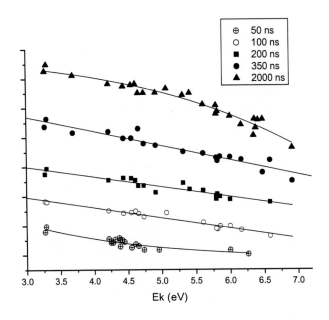

Figure 3.2. Boltzmann plot measured over Fe I lines at different delays from the laser pulse; the sample is stainless still in air.

In the early plasma stage, the ionic temperature is higher than the atomic one, and by measuring the both of them, an indication about LTE existence can be obtained. In Figure 3.3 there is an example of the temperature measurements through the Boltzmann plot, in the case of a rock sample analyzed in air and at low pressure environment (6 mbar of CO_2) [Colao et al 2004]. In order to check the presence of Boltzmann equilibrium over atomic levels, the temperature of Fe I was calculated both over all the emission lines free of overlap (T_{exc}) and for the transitions with higher energy levels Ek > 4.5 eV (T_{high}). The ionic temperature was measured on Ti II emission lines, but due to a low E_k energy difference (3.8-5.8 eV) these measurements must be considered with a large error. As expected, in early phase of plasma the calculated ionic temperature T_{ion} is higher than the atomic. Initially, also the atomic temperature T_{high}, closer to equilibrium with lower ionic levels, is higher than the temperature T_{exc} measured over all the energy levels. The overpopulation of the lower levels (Figure 3.2, delay 50 ns) is characteristic for the ionizing plasma [Van der Mullen 1990]. After the initial fast plasma cooling, all the three temperatures here measured, tends towards the same value. In air atmosphere, all the three measured temperatures could be considered equal for delays 600-3000 ns (Figure 3.3a). In low-pressure CO_2 atmosphere, the three temperatures achieve approximately the same values after 200 ns. However, T_{high} shows a significant increase already after delay of 1000 ns (Figure 3.3b), where the electron density in highly expanded plasma becomes lower than value require for LTE (see eqn. 3.4). For the described measurements at low pressure surrounding the LTE could be valid only for delays 200-1000 ns. This interval is even more restricted for lower laser energies, and by applying 7 mJ instead of 14 mJ, it results limited to delays 100-500 ns [Colao et al. 2004].

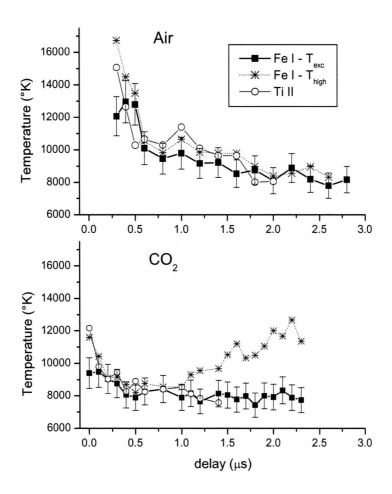

Figure 3.3. Plasma temperature in air (atmospheric pressure) and in CO_2 (6 mbar), measured over all Fe I lines (T_{exc}), over Fe lines with $E_k > 4.5$ eV (T_{high}) and for Ti II lines (E_k 3.8-5.8 eV) as a function of delay; acquisition gate 100 ns, laser energy 14 mJ.

3.3. Electron Density

Once the plasma temperature is known, the electron density can be calculated from the spectrum: one way to calculate it is through the Saha equation (Equation 3.3). Indeed Equation 3.2 allows for retrieval of atomic and its first ionization degree concentrations N_α^I and N_α^{II}, from the line intensities. Then the electron density is derived from Equation 3.7:

$$N_e = \frac{N_\alpha^I}{N_\alpha^{II}} \frac{U_{II}(T)}{U_I(T)} B(kT)^{3/2} e^{-\frac{E_\infty}{kT}} \tag{3.7}$$

These calculations require the existence of LTE over a large span of the level energies, which includes both the atomic and ionic levels.

Under usual LIBS conditions, the line widths are mainly determined by the instrument and by the Stark broadening. The latter is caused primarily by collisions with the free electrons, although there is a smaller contribution of the collisions with ions. The most common method for determining of the electron density in LIBS plasmas is based on measurement of the Stark broadening relative to the atomic or ionic lines for which the broadening coefficients are known in the literature [Griem 1974].

Except for hydrogen lines, for which the formula is slightly different, the electron density could be calculated from the measured Stark width $\Delta\lambda_s$ of the line [Konjevic 1999]:

$$\Delta\lambda_s \cong 2w(T) \cdot [1 + 1.75 \cdot 10^{-4} N_e^{1/4} A(T)(1 - 0.068 \cdot N_e^{1/6} T^{-1/2})] \cdot 10^{-16} N_e \tag{3.8}$$

where $w(T)$ is electron impact half-width, N_e is electron density (cm^{-3}) and $A(T)$ is ion-broadening parameter. The second and the third term in the brackets can be neglected in the case of LIBS plasmas, so the expression from above is simplified and rewritten as:

$$N_e \cong \frac{\Delta\lambda_s}{2w(T)} \cdot 10^{16} \tag{3.8}$$

If the other types of line broadening, such as Doppler, Wan der Waals and resonant broadening can be neglected [Konjevic 1999], as common for the LIBS plasmas [Colao et al. 2004], the Stark broadening can be obtained from deconvolution of the line. In general case, the line profile is described by Voigt function, where the Gaussian component is due to the instrumental broadening (experimentally determined) and remaining FWHM is attributed to the Stark broadening described by Lorentz's distribution. For so determined Stark width, the electron density is calculated from Equation 3.8 by introducing the tabulated coefficients for the given line [Griem 1974], previously extrapolated for the measured plasma temperature.

3.4. Self-Absorption

A plasma is optically thin when the emitted radiation traverses and escapes from the plasma without significant absorption. In the case of strongly emitting atomic lines from the main sample constituents, as for example some lines of calcium and sodium in soils, the effect of re-absorption cannot be neglected i.e. the approximation of optically thin plasma is not applicable. As a consequence, the relationship between the integrated line intensity and the species concentration (Equation 3.2) is not valid. This also leads to a non-linear behavior of the calibration curves (see section 6), showing a saturation at high element concentrations.

Optical thickness at a certain transition wavelength causes the line broadening due to self-absorption, and sometimes, even the line self-reversal (Figure 3.4). In the case of multiplet transitions, the existence of self-absorption for these lines can be checked from their intensity ratios. In the presence of the line absorption, the stronger lines when compared to those less intense from the multiplet have lower values than those given in the data-bases.

For optically thick plasma at a certain strong transition, the line intensity variation at frequency ν, through the plasma layer dl, is given by [Lochte-Holtgraven 1995]:

$$dI_\alpha^{ki} = (\varepsilon_\nu^{ki} - k_\nu^{ki} I_\alpha^{ki})dl \qquad (3.9)$$

Where ε_ν and k_ν are coefficients of plasma emission and absorption, which depend on the emission frequency, original line intensity distribution $P(\nu)$, Einstein coefficients (A_{ki}, B_{ki} and B_{ik}), and number of species at upper and lower level respectively:

$$\varepsilon_\nu^{ki} = \frac{A_{ki}}{4\pi} N_k h \nu P(\nu) \qquad (3.10)$$

$$k_\nu^{ki} = \frac{h\nu}{c}(B_{ik}N_i - B_{ki}N_k)P(\nu) \qquad (3.11)$$

where we neglect the absorption due to Bremmstrahlung and to the recombination.

From equations (3.9-3.11), assuming intensity equal to zero at the sample surface and a Boltzmann population distribution of the energy levels, the line intensity at the frequency ν that is escaping from homogeneous plasma along a line of sight of length L, is given by:

$$I_\alpha^{ki}(\nu) = \frac{2h\nu^3}{c^2}\frac{1}{1-e^{-h\nu/kT}}(1-e^{-k(\nu)L}) \qquad (3.12)$$

This equation is applicable as long as a stimulated emission from the upper level can be neglected [Lochte-Holtgraven 1995]. In order to obtain the overall line intensity, it is necessary to integrate equation (3.12) and this can be performed only numerically in the case of the Voigt profile. The numerically integrated intensity can be correlated to the number of the single species in the plasma (i.e., the species concentration) through the absorption coefficient (equation 3.11) and Boltzmann distribution over all energy levels. The application of equation (3.12) to LIBS concentration measurements relevant to strongly absorbing lines does not result easily practicable, requiring the knowledge of the initial line profile as well as laborious numeric calculations [Aragón et al. 2005].

Self-absorption of the lines reduces the accuracy in the plasma parameters estimation, and a careful choice of the emission lines for such calculation must be performed [Aragón et al. 2005, Colao et al. 2004].

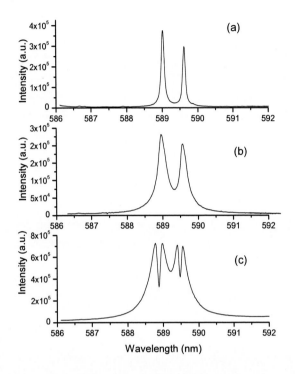

Figure 3.4. Na resonant lines detected by LIBS: (a) optically thin plasma at these transitions; (b) in presence of self-absorption; (c) self-reversal due to a strong self-absorption.

4. QUALITATIVE ANALYSES

Qualitative analyses are aimed to establish the elements present in the sample, by exploiting the information present in the registered spectrum. This information includes: the wavelength of the emission lines, their intensities and relative intensities of the lines belonging to a single element at a certain ionization stage. The main problem in LIBS analyses of unknown samples is related to the overlap of the emission lines from different elements, both atomic and ionic, given in the data bases.

4.1. Line Assignment

In order to assign the emission lines from the LIBS spectrum and thus to determine the elements present in the sample, it is necessary to consider:

- The LIBS spectra, in typical time detection interval, have the detectable lines only from the atoms and single ionized species. Emission from higher ionization stages can be generally neglected.

- Presence of the ions and their emission intensities are strongly dependent on the plasma temperature, which is higher for shorter acquisition delays from the laser pulse (section 3.2, Figure 3.3). Higher plasma temperature means a higher fraction of the ions with respect to the atoms, and consequently their more intense emission (Figure 4.1).

- When using an acquisition delays in order of some microseconds the ionic emission becomes rather weak and the spectrum is dominated by the atomic lines, with the exception of some strong ionic lines (as for example – resonant lines) belonging to the elements with a low ionization energy. In such conditions, if an ionic line from one element overlaps with some atomic emission from another element that can be present in the sample, it is more probable that the line belong to the atomic emission.

- Dynamic of the emission lines also depend on the composition and pressure of surrounding atmosphere (Figure 4.1, see also section 5.1).

- *A priori* knowledge of the sample type is very useful for assignment of the emission lines. As for example, when analyzing a steel sample, beside Fe, one can expect to detect also Ni, Cr, Mo, C etc., while Ag, Au, Pb, U, etc are not expected. So, the lines of the elements probably not present in the sample can be excluded anytime they overlap with the lines of certainly present elements.

- Relative line intensities given in the databases, although they can differ from one database to another, can be exploited for the line assignment. Particularly, the multiplet lines of one element are useful for the identification – if one of them is present and relatively strong (quiet above the detection threshold), also the others must be observed. In such case the element identification is straighford.

Let's see one example:

If we observe an emission line close to 390.5 nm on some completely unknown sample, from the data base it could be seen that there is a number of the tabulated lines [NIST] close to this wavelength (Tab, 4.1). The width of the spectral range around this position that should be considered, depends on the resolution of the detection system and accuracy in the initial wavelength calibration. If we consider the range of 390.50 ± 0.05 nm and exclude the presence of highly ionized species (example Fe VI), still there are different elements that could be assigned to this line: Th, Er, Si, Cl, Gd, Te, Ho and Nd. Particularly, the rare earth elements here have the relatively strong tabulated line intensities. However, these elements are in traces for most of the sample types. If we consider soils, the most probable element to be assigned to this line would be Si. However, in order to be sure that this emission really belongs to Si it is opportune to check the presence of a relatively strong Si I multiplet close to 251 nm (Table 4.2). If these lines, or at least the most strong among them are also present in the spectrum and with the similar relative intensities, the single line from above can be assigned to Si. Otherwise, the presence of other relatively strong lines in the spectrum belonging to other elements given in Table 4.2 should be checked in the other spectral parts.

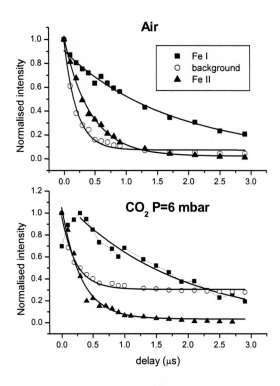

Figure 4.1. Time evolution of the plasma background (continuum) emission and of the integral line intensities for Fe I and Fe II lines in air and CO_2 atmosphere; the sample is a soil, the acquisition gate is 100 ns [Colao et al 2004].

Table 4.1. Emission lines from NIST data base [NIST] close to Si I line at 390.5 nm

Species	Wavelength [A]	Relative Intensity
Sc I	3904.657	5
Ti I	3904.78	2600
Yb II	3904.81	50
P III	3904.811	200
Fe VI	3905.01	-
Fe VI	3905.01	-
Th II	3905.186	170
Er I	3905.4	1200
Si I	3905.521	300
Cl II	3905.621	-
Gd I	3905.65	450
Te II	3905.67	20
Ho II	3905.68	1300
Nd II	3905.89	1700
Fe II	3906.035	-
Bk II	3906.09	10000
Kr II	3906.177	150
Co I	3906.285	-

Er II	3906.31	11000
Hg I	3906.37	60
Sc I	3906.432	4
U I	3906.45	380
Fe I	3906.479	250

Table 4.2. Silica multiplet lines corresponding to the resonant transitions from the upper state with the energy of about 4.92 Ev

Species	Wavelength [A]	Relative Intensity
Si I	2506.897	425
Si I	2514.316	375
Si I	2516.112	500
Si I	2519.202	350
Si I	2524.108	425
Si I	2528.508	450

4.2. Sample Identification

Through the element identification in the sample, also different type of the materials can be recognized. This is for example important, also more difficult, in the case of the archeological findings of unknown origin. While the main constituents determine the class of the materials, presence of some trace elements could even supply information about the manufacturing procedure and period, sometimes also about the geographical origin of the finding. In Figs. 4.2-4.4 the examples of the LIBS spectrum obtained on different classes of the materials are shown [Lazic et al 2005b], namely:

- *iron alloy*: (commercial alloy C40), containing Manganese (0.5-0.8%), Carbon (0.37-0.44%), then traces of Chromium (<0.1%), Molybdenum (<0.1%) and Nickel (<0.1%).
- *bronze alloy*: main constituents of the bronzes are Cu (here 82.47%), Sn (here 5.29%), Pb (here 5.55%) and Zn (here 5.86%)
- *silver alloy*: a commercial silver 900/1000 was used having 90% of Ag and 10% of Ni.
- *gold alloy*: a commercial gold 18 K containing Gold concentration of 75% and 25% of unknown elements.
- *marble*: the fragment was collected in one quarry utilized in ancient times and whose composition was determined in our previous work [3d] (Ca 40.8%, C 12.2 %, Al 1.20%, Si 0.72%, Mg 6940 ppm, Fe 486 ppm, Ti 366 ppm, Cu 250 ppm, Mn 165 ppm, Ba 98 ppm).
- *calcareous rock,* extracted from a sea bed

The differences in the spectral features are evident even in a narrow spectral range (270-290 nm), and though the line assignment it was possible to identify the main sample constituent and consequently, the type of the material (Table 4.3).

From the spectra obtained on the iron alloy, in addition to a number of Fe I and Fe II lines (Figure 4.2a), also Mn was detected from its ionic triplet around 260 nm and its atomic emissions (Table 4.3), and then Cr ionic (Figure 4.2a) and atomic emission around 425 nm (Table 4.3). On the contrary, C, Mo and Ni concentrations resulted below the detection limit in this experiment.

All the four main elements forming the bronze alloy were identified in the same spectral range from above (270-290 nm). However, the strongest Zn emission was observed around 470 nm (Table 4.3) and this region should be used for the detection of Zn at low concentrations. The plasma emission was sufficiently strong to detect also some minor, non-certified elements, such as Ni and Fe (Table 4.3).

The spectra of the precious alloys were intense although acquired by applying a single-shot, aimed to avoid any visible damage on the surface (Figure 4.3). On the silver sample, different atomic and ionic emission lines from Ag and Ni were detected (Figure 4.3a). The presence of other elements could not be inferred from the spectra acquired also in other wavelength intervals. The examined gold alloy had the composition certified only for gold (75% wt), while the other elements were unknown. Typically, Ag and Cu are added to the gold alloys used in jewelry (the sample was a bracelet), and frequently these alloys also contain Zn and/or Ni. Co or Fe could be also present as traces, as well as Ir or Rh, whose addition reduces the dimensions of the alloy grains. From the UV-VIS spectra, Au, Ag, Cu and Zn (Figure 4.3b) were detected, the first three elements both from atomic and ionic emission lines. All the mentioned elements had emission line intensities largely above the detection threshold in the spectral range 270-290 nm.

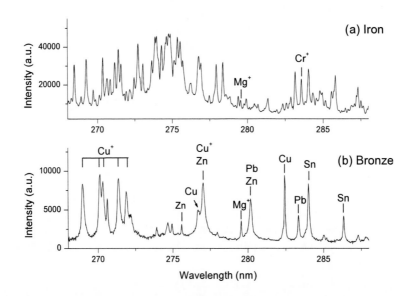

Figure 4.2. (a) Iron sample C40: detection of iron atomic and ionic lines (unmarked peaks) and chromium ion; (b) bronze sample: detection of Cu, Pb, Sn and Zn.

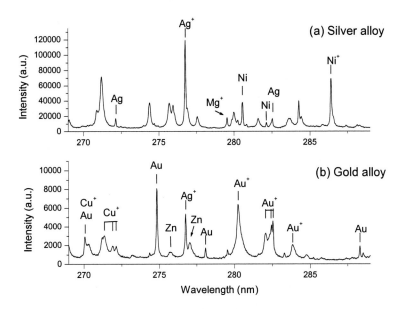

Figure 4.3. Recognition of precious alloys: (a) Silver alloy: atomic and ionic emission from Ni and Ag; (b) Gold alloy: detection of gold (Au and Au$^+$), Copper (Cu$^+$), Zn atomic and Ag ionic emission.

Let's see how to distinguish two materials with similar matrices. For both marble and calcareous rock the expected main constituent is $CaCO_3$, however, a larger amount of impurities could be expected in the calcareous mineral. In here discussed case the data files were registered for 50 successive laser shots. The spectral intensities were much weaker for the clean, white marble than for the darker calcareous mineral, the difference being probably related to the better optical coupling of the near infrared laser radiation with the surface of the latter. The spectral signature from marble corresponds to its major constituents: Ca, C and Mg. Traces of Si and Fe (Figure 4.4a, Table 4.3), Al and Sr (Table 4.3) were also detected after applying the signal filtering [Lazic et al. 2005]. Note that Mg, and to a smaller extent Sr, are common calcium substitutes in calcium carbonate matrices.

The spectra from the rock sample showed the same main elements as the marble, but with more intense emissions from Al and Sr. Also emissions from Si and Fe were significantly stronger than in the case of the marble and more lines from these elements were observed (Figure 4.4 b, Table 4.3). However, several other impurities were clearly identified in the rock samples, such as Mn and Ba . While Si and Al are characteristic of alumino-silicates, not expected in the marbles, Mn, Fe and Ba are usually present in the sedimentary rocks and are related to the past biological activity in seawater where the rock had been collected. LIBS analyses of marine sediments and a comprehensible interpretation of the presence of different elements have been already reported in our previous work [Barbini et al. 2002].

Although these measurements were performed underwater, analogue procedure fir the qualitative analyzes could be also applied in air surrounding.

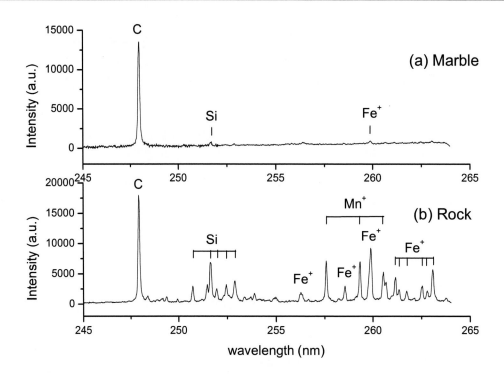

Figure 4.4. Comparison of LIBS spectra from marble (a) and calcareous rock (b) in the spectral range 245 – 265 nm: identification of C, Si, Fe and Mn.

Table 4.3. Element transition lines or bands (wavelength in nm) used for the material recognition.

Elements	Material					
	Iron	Bronze	Gold alloy	Silver alloy	Marble	Rock
Fe	Different	$260^{+(b)}$	-	-	259.9^+	$260^{+(b)}$, $275^{+(b)}$, 302.5
Cr	$283.6^+, 425^{(b)}$	-	-	-	-	-
Cu	-	$270^{+(b)}$, $282.4, 465.1$	$270^{+(b)}$, 465.1	-	-	-
Pb	-	283.3	-	-	-	-
Sn	-	$284.0, 286.3$	-	-	-	-
Zn	-	275.6, $468.0, 472.2$	275.6, 277.1, $468.0, 472.2$	-	-	-
Ag	-	-	276.8^+, 338.3	276.8^+, $272.2, 282.5$	-	-
Au	-	-	274.8, 280.2^+, $282^{+(b)}$	-	-	-
Ni	-	$338.0, 341.5$	-	280.5, $282.1,$	-	-

					286.4^+	
C	-	-	-	-	247.8	247.8
Ca	-	-	-	-	$300^{(b)}, 458^{(b)}$	$300^{(b)}, 458^{(b)}$
Mg	-	-	-	-	$280^{+(b)}$, 284.8	$280^{+(b)}$, 284.8
Si	-	-	-	-	251.6	$252^{(b)}$, 288.1
Al	-	-	-	-	308.2,309.3	308.2, 309.3
Mn	257.8^+, 259.3^+, 260.6^+, $476^{(b)}$, 478.3	-	-	-		257.8^+, 259.3^+, 260.6^+, $295^{+(b)}$
Ba	-	-	-	-	-	455.4^+
Sr	-	-	-	-	460.7	460.7

(b) = emission band around the specified wavelength; (+) = ionic emission

4.3. Depth Profiling

LIBS is essentially a surface analysis technique, as a plasma is produced by ablation of the top surface layer, typically less then 1 μm. On an approximately homogeneous material, the signal can be accumulated after repeated pulses at the same point in order to achieve a better SNR. Such sampling can be also used on inhomogeneous samples with aim to average the spectra i.e. the sample depth composition.

As the each laser shot, if sent at a fixed point, removes a certain thickness of the surface layer, this can be exploited to clean initially the surface and then to proceed by the LIBS analyses on the layers of interest. For example, if a rock is covered with a soil or dust, first laser shots can be used to remove unwanted superficial material and then to proceed with the bulk rock analysis.

By registering the spectra after each laser shot at a fixed point, it is possible to analyze the depth profile of the elements in the sample and up to a few millimeters. Knowing the ablation rate of the material, which must be determined experimentally, the element distribution in a small scale can be so estimated. The depth profiling by LIBS is also used for the process monitoring during the laser cleaning. Each laser pulse removes a certain thickness of the material and also produces the plasma. By monitoring the plasma emission, it is possible to establish when the unwanted surface layer, whose composition is different from the underlying material, is removed.

As for example, the LIBS was used to monitor removal of the encrustation from marble surface [Lazic et al 2004], with aim to determine the right point when the process must be interrupted. Initially, both the crust and clean marble sections were fully analyzed by LIBS. The crusts here considered show a relative abundance of Si, Al, Ti, Fe and Mn, but their concentrations decrease rapidly with the depth down to the levels present in the bulk (Figure 4.5). Based on the depth distribution of these elements, the crust could be considered removed with less than 50 laser shots, after which the surface composition remains practically constant, thus indicating that the bulk material has been already reached. To estimate the ablation rate, a thin layer of bulk sample has been cut and the number of laser shots necessary to drill a hole in it was measured. The ablation rates on the considered samples were between

1.0 and 1.5 μm per shot. These values averaged over the crust depths and although could differ significantly from one crust layer to another, they are useful particularly when a large number of the laser shots is applied for averaging. In this work, the element concentrations were measured by LIBS after generation of the calibration graphs and implementation of an appropriate data correction procedure (section 5). However, in most of the cases, simple measuring the element line intensities or their ratios can monitor the cleaning process.

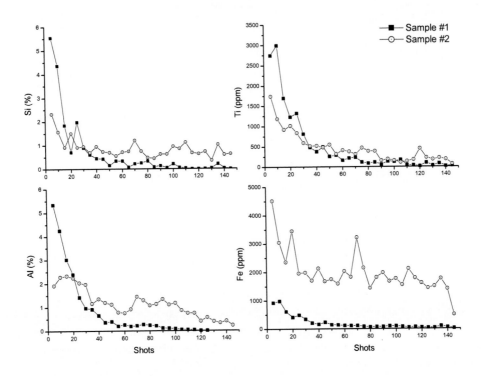

Figure 4.5. LIBS measured concentration change of Si, Al, Ti, Fe and Mn during the encrustation removal from two different marble samples.

5. QUANTITATIVE ANALYSIS

Quantitative analyses are aimed to determine the concentrations of different elements in the sample. LIBS technique applied for measurements of the element concentrations is not considered a very accurate and highly sensible technique as for example ICP-OES. Typically, LIBS analyses of a-solid samples with a-priori known matrix have uncertainty 5-10%. The detection limits are in ppm range and depend on the element analyzed. The analytical accuracy strongly depends on the experimental conditions, where some of the parameters, such as the laser energy fluctuations, can be controlled and corrected. Systematic errors can also be originated by the sample physical and chemical properties [Wisbrun et al. 19994, Anzano et al. 2006]. The latter is related to the so-called matrix effect [Boumans 1976, Eppler et al. 1996] i.e. to the strong influence of the material composition on the produced plasma emission and therefore on the final analytical results.

5.1. Variables Affecting the Quantitative Analyses

5.5.1. Sample Properties

Sample overall chemical composition has a great influence on the relative emission line intensities in the plasma. If the sample contains a high amount of easily ionized elements, such as alkali elements, the laser produced plasma has a higher density of the electrons. The increase of the electron density leads to a more efficient electron-ion recombination. As a result, the ratio of neutral to ionic species in the plasma increases, as it can be seen from the Saha equation (Equation 3.3), and this affects the concentration measurements [Eppler et al. 1994].

In addition to the chemical composition, the physical properties of the sample, such as: absorption at the laser wavelength, heat capacity, vaporization temperature, grain size and material compactness (some of these properties are chemically dependent), influences the laser ablation rate and also the plasma parameters. The grain size is related to the persistence of the laser generated aerosol above the sample surface, with consequences for the signal intensity (see the next section). Mixed effects of the grain dependent ablation rate and aerosol density, can bring to an apparently contradictory results. In [Carranza et al. 2002] the LIBS measurements were performed on powder samples and aerosols, concluding that smaller size of the particles leads to more effective plasma formation and consequently, to more intense line emission. Opposite effect was observed in [Wisbrun et al. 1994], where the signal from the soil contaminants results proportional to the grain diameter. As a result, different slopes of the calibration curves were obtained for the samples with different grain sizes.

In [Lazic et al. 2004] the LIBS spectra of marbles (very compact material), which have $CaCO_3$ matrix and some impurities, are compared with the spectra of $CaCO_3$ powders pressed into pellets and containing the similar impurities (added). It was noticed that the overall spectral intensity increases with the $CaCO_3$ content, both when analyzing the pellets and marbles. However, the samples prepared by powder pressing exhibit a significantly stronger plasma emission than natural marbles (Figure 5.1), which can be related to the different ablation rates. Also, the plasma parameters (section 3), important for accurate LIBS quantitative analyses, differ strongly from one sample to another (Table 5.2).

5.1.2. Sampling Geometry and Laser Energy Density

Changes in the geometry of the laser focusing and signal collection can strongly affect the LIBS spectra. The laser energy density on the sample surface, which is determinant for the ablation rate per pulse and for the signal intensity as well, depends on the focal length. If this one is shorter, the system is more sensible to defocusing i.e. to the variations of the Distance Lens-Surface (DLS). This factor becomes critical when analyzing the rough surfaces. With the changes of the energy density on the sample, both the plasma shape and the plasma parameters changes [Yalcin et al. 1999], as well as a fraction of the atomized material with respect to the material removed mechanically from the surface by shock-waves end eventual splashing. The optimal sample position for obtaining the highest signal by laser-induced ablation of solids is slightly above the focal plane [Bulatov et al. 1996, Multari et al. 1996].

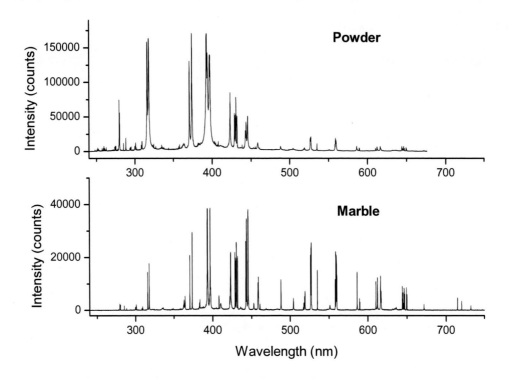

Figure 5.1. LIBS spectra of a CaCO₃ powder pressed into a pellet and of one clean marble fragment: all strong lines belong to Ca I and Ca II while the emission from other species have intensities lower than 5% of the maximum scale.

5.1.2. Sampling Geometry and Laser Energy Density

Changes in the geometry of the laser focusing and signal collection can strongly affect the LIBS spectra. The laser energy density on the sample surface, which is determinant for the ablation rate per pulse and for the signal intensity as well, depends on the focal length. If this one is shorter, the system is more sensible to defocusing i.e. to the variations of the Distance Lens-Surface (DLS). This factor becomes critical when analyzing the rough surfaces. With the changes of the energy density on the sample, both the plasma shape and the plasma parameters changes [Yalcin et al. 1999], as well as a fraction of the atomized material with respect to the material removed mechanically from the surface by shock-waves end eventual splashing. The optimal sample position for obtaining the highest signal by laser-induced ablation of solids is slightly above the focal plane [Bulatov et al. 1996, Multari et al. 1996].

The laser ablation also produces an aerosol, composed of substrate material. The aerosol persistence above the sample surface depends on the ambient pressure (see section 5.1.4). When a dense aerosol formed by the previous laser pulse persists and relative long focal lengths are used, the energy density needed for plasma formation can be reached already much above of the analyzed surface [Bulatov et al. 1996]. The plasma formed far away from the sample causes a partial absorption of the laser energy before reaching the sample. Consequently the sample ablation rate and the LIBS signal, is considerably reduced. On the other hand, the same aerosol presence, although scatters the incoming laser radiation, can contribute to a signal increase for intermediate focusing distances. Here, the threshold for plasma formation on the aerosol is reached close to the sample surface, in the volume occupied by the plasma produced by the sample ablation. In this case, the aerosol containing the target particles is excited in the hot plasma and can even increase the intensity of the analytical lines as the sample material is already available without spending the laser energy on surface ablation.

For a fixed energy density on the sample, the ablation rate (so the LIBS signal) changes with the defocusing, and on a different way when using a spherical or a cylindrical lens [Multari et al. 1996]. In the case of an elliptical or rectangular spot on the sample, both the ablation rate and the LIBS signal are dependent on the spot aspect ratio [Bulatov et al. 1996].

Regarding the optical collection system, the signal is influenced by a portion of the plasma imaged onto the optical fiber and/or the spectrometer entrance slit. With decreasing of the surrounding gas pressure, the plasma becomes more expanded and it is more difficult to collect the signal from the whole plasma volume. As for example, in air at the atmospheric pressure, the produced plasma can have a diameter of only 1-2 mm, while in vacuum, it expands to the visible diameter of more than 1 cm.

Other important factors are – the plasma shape and which portion of the plasma radiation is collected by the system. The plasma at atmospheric pressures tends to elongate towards the focusing lens with increase of the laser energy [Yalcin et al. 1999]. If the collection system collinear with the focusing one (mono-axial configuration), a great amount of the emission generated closer to the sample surface can be lost due to absorption by the plasma volume nearer to the focusing lens.

The plasma also has a temperature distribution, with a hot core and colder external regions [Aguilera et al. 2004]. The latter contribute particularly to the self-absorption of the resonant emission lines. The species density distribution is not uniform [Bulatov et al. 1996], and there is a higher fraction of atoms in the plasma core [Aguilera et al. 2003, Castle et al. 1997] due to higher electron density and more efficient recombination. Also, the ions are generated in the earlier plasma stage, and by the plasma expansion, brought before the atoms to the plasma edges [Siegel et al. 2005]. All this determine a geometry dependent intensity of different emission lines even belonging to a single species [Castle et al. 1997].

In Figure 5.1, an influence of the laser energy at fixed DLS on the signal temporal behaviour is shown for two different gas atmospheres [Colao et al. 2004]. The measurements were performed on a rock sample. The signal was considered as fitted peak intensity after the background subtraction, while the noise is given by the standard deviation of the background emission close to the chosen line with approximately the same spectral width. In both gaseous environments the SNR has a maximum at a certain gate delay from the laser pulse. After this maximum, the signal decays, more slowly when higher laser energies are applied and this makes the choice of the optimal gate width less critical. In air, the SNR values and temporal evolution are quite similar when applying 21 mJ and 14 mJ, so the use of lower laser energy here allows for reducing the system power consumption without significant signal degradation. This effect was already observed by other authors [Wisbrun et al. 1994]: the threshold for the laser energy after which a further energy increase does not produce any analytical improvement does depend on the overall experimental conditions and sample properties. Differently, in CO_2 at reduced pressure the maximum SNR value is achieved at earlier plasma stage and with a smaller peak when the laser energy is reduced. In addition, the SNR decay becomes faster at lower laser energies, which requires shorter gate widths in order to keep high SNR, particularly in conditions of an intense background illumination. The optimization of the experimental parameters, aimed to obtain a high SNR ratio is particularly important for detection of trace elements. However, the variation of SNR with delay and its peak position on the time scale also depends on the choice of the analytical line [Aragon et al. 1999].

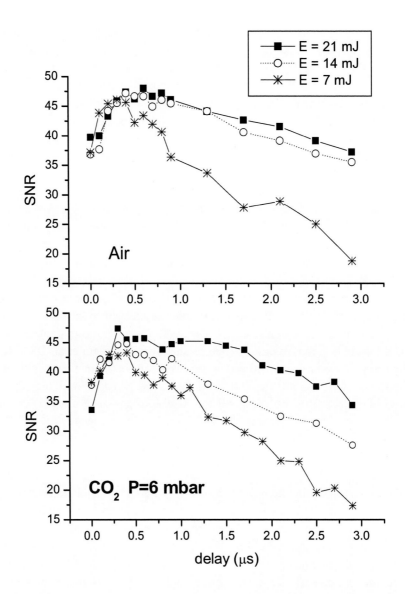

Figure 5.2. Signal-to-noise ratio (SNR) of Fe I line at 376.7 nm, as a function of the acquisition delay from the laser pulse in the air and CO_2 atmosphere; pulse energies are 7 mJ, 14 mJ and 21 mJ; gate 100 ns.

5.1.3. Sampling Method

As discussed before, the laser ablation produces a persistent aerosol above the sample. An aerosol amount increases with the laser repetition rate and this can even enhance the LIBS signal by supplying already ablated material into the hot plasma. Nevertheless, presence of aerosols beyond a certain concentration reduces the signal, both due to increased laser scattering and shielding by the plasma formed before reaching the sample surface. At fixed laser energy, the optimal repetition rate is strongly matrix dependent. This optimization was performed for soils and sands [Wisbrun et al. 1994], and in the given experimental condition it corresponds to 1 Hz and 6 Hz respectively. These differences indicate an additional matrix effect with respect to the chemical one (section 5.1.1.). The same authors reported that the relative standard deviation of LIBS signal is low up to a repetition rate of 8 Hz and then increases dramatically.

The laser ablation creates small craters on the surface by laser-induced evaporation and material removal by the shock waves produced. On the compact materials such as metals, glasses, rocks etc., the crater diameter is close to the laser beam diameter, while the depth depends on the ablation rate and on the number of the applied shots. When the laser pulses are applied to soft materials such as soils, the laser produced craters are much larger, up to a few millimetres in depth and diameter, because the material particles are easily removed by the shock waves. For sampling at a fixed position, a crater depth varies DLS and the distance of the plasma plume to the collecting optics. The plasma parameters also changes with the crater depth as the laterally confined plasma is denser and hotter than the plasma expanding semi-spherically into a free space [Sallé et al. 2005, Mao et al. 2004].

Beside the crater effect and aerosol formation, which influence the LIBS signal, there is also a possible local surface enrichment with relatively large particles and heavier elements. This happens since the lighter particles are removed much easier by the shock wave. Then, the successive laser pulses find the surface containing different average particle size and with a different composition form the target (section 5.1.1.).

All here mentioned factors, matrix-dependent, affect the measuring accuracy. However, they can be partially overcome by sampling at different, always fresh points [Wisbrun et al. 1994]. This is practicable by moving the sample (example rotation) or the optical head containing both the focusing and collecting system. The aerosol itself can be removed by a slight gas flow above the solid sample.

5.1.4. Surrounding Environment

The LIBS plasma is produced inside gaseous or liquid environment, so its composition and pressure influences the plasma characteristics and the results of the measurement [Sdorra et al. 1992].

Reducing the atmosphere pressure, the plasma generated at the beginning of the laser pulse is less dense as it expands more effectively into a low pressure surrounding. In case of nanosecond laser excitation, the remaining (later) portion of the laser pulse is then less shielded by the plasma itself, i.e. better coupled to the target. This leads to an increase of the ablation rate with reducing of the surrounding gas pressure down to a few mbar. The fraction of the atoms in plasma decreases with lowering of the pressure due to reduced number of collisions in more expanded, less dense plasma. Both the plasma excitation and decay rate are determined by the collisions, which are pressure dependent. The overall influence of the gas pressure on the LIBS signal intensity is a result of he above mentioned factors and of the changes in the plasma geometry. Increasing the ambient pressure the plume has smaller dimensions, and then the optical collection of plasma emission is more efficient.

Ionization potential of the ambient gas is of a particular importance for the LIBS signal. A low ionization potential means that is easier to produce the plasma and that its temperature is expected to be higher. Another effect is related to the cooling processes of the excited analyzed atoms/ions in the plasma, which is more efficient for lighter gases (example He) and this leads to a faster plasma decay. In summary, the measured LIBS intensities are higher as the surrounding gas is heavier and with low ionization potential [Wisbrun et al. 1994]. For example, the analysis in Ar at the atmospheric pressure can increases the signal by a factor of 1.8 with respect to the air surrounding at the same pressure.

Another interesting effect was observed when generating the calibration graphs on different certified soil matrices (soil and clay samples) [Sallé et al. 2005]. At the pressure of 585 Torr, the slope of the calibration curves was quite different due to the matrix effect. Lowering the pressure to 50 mTorr, these curves overlap so indicating a matrix effect reduction.

5.2. Different Approaches in Quantitative LIBS Analysis

Common approach for obtaining quantitative material analyses by LIBS is based on use of the calibration curves generated after measurements on the reference samples. The main limitation of this approach is related to the matrix effect i.e. strong influence of the material composition and its physical properties on the produced plasma emission and on the final analytical results (see section 5.1.1.). Due to this problem, the initial calibration should be performed with standards having the matrices similar to that one of the samples to be characterized, and reliable concentration values can not be retrieved by LIBS on a priori unknown samples. Even for the well matched standard matrices, different sources of analytical errors could be present, such as slight variation of the experimental conditions and differences in the coupling of the laser radiation with the samples, as well as different grain size in the case of the soft samples.

In order to improve LIBS analytical accuracy by reducing the effects of variable experimental conditions and ablation rates, different kinds of the LIBS signal normalization have been proposed. The examples include lines intensities normalization: on the acoustic plasma emission [Chaleard et al. 1997], on the continuum plasma emission [Lazic et al. 2004, Body et al. 2001], on the overall plasma emission [Lazic et al. 2001] and on a line intensity of some major matrix element [St-Onge et al. 1997]. By the latter type of normalization only the relative concentrations can be retrieved, except for well characterized matrices with a fixed content of the element used for the normalization. In Figure 5.3 there is an example of the calibration graph obtained before and after the line intensity normalization on the integrated plasma emission [Lazic et al. 2001]. In the latter case, the point scattering on the calibration graph is significantly reduced, as it can be seen from the higher correlation factor R.

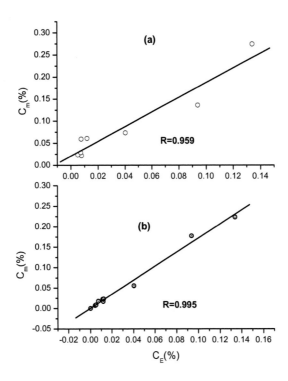

Figure 5.3. LIBS measured vs. certified Cr concentration for different soil-like reference samples (atomic transition at 425.43 nm) before (a) and after the line intensity normalization (b) on the integrated spectral emission (240-750 nm).

In [Lazic et al. 2004] the intensity of each analytical line, measured on marble samples and $CaCO_3$ powders, was firstly normalized on nearby background emission. Here, it was hypothesized that the continuum plasma emission is correlated to the ablation rate through a complex equation including electron and ion densities in plasma [Griem 1964]. This correlation was checked by measurements of the ablated volume on one marble sample by applying 5, 10 and 20 laser shots at two different pulse energies (12 and 20 mJ). The marble surface was lapped before applying the laser pulses. The crater volumes were determined by a standard profilometer and averaged over five measurements. The intensity of plasma continuum acquired during the ablation was averaged in the interval 290–293 nm. In spite of large difference between the two laser energies used, the relation between the continuum plasma emission and the ablated volume was linear and well correlated (R ~ 0.992) as shown in Figure 5.4. This fact supports the use of the line intensity normalization on the background emission as a tool for minimizing the influence of the ablation rate variability on the analytical results, which could originate from differences in physical-chemical sample properties, from laser instabilities and from an imperfect focusing. Examples of the calibration graphs obtained for different elements after the normalization on the plasma continuum are shown in Figure 5.5. The element lines selected for the calibration and the obtained correlation factors are given in Table 5.1. The analytical line intensities refer to the line integrals obtained through the line fittings by Lorenz functions, which describes well the line shape in high density plasmas as here. The use of the line integrals instead of the peak values, together with the normalization to continuum emission intensity, reduced significantly the data scattering on the calibration plots. For Na sodium line at 589.99 nm and the carbon line at 247.86 nm (calibration plots not reported), the correlation coefficients were lower than for other elements and a linear fit does not pass through the origin (0,0) of the axes. In the first case, the poor correlation could be attributed to sodium variation in the atmosphere, as expected for the proximity of our laboratory site to the coast. The line intercept at positive Y-axis value is attributed to always present, interfering sodium from air. Regarding carbon, the interference of the nearby iron spectrum is an important source of error in the line intensity determination. For all analytical lines calibration data were fitted by straight lines except for Si I and Mg I lines, where the nonlinear fitting functions were used to describe the data saturation [Lazic et al. 2001].

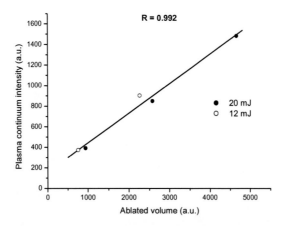

Figure 5.4. Intensity of plasma continuum for spectra acquired with 5, 10 and 20 laser pulses of energies of 12 and 20 mJ, as a function of the marble volume ablated with the corresponding shot numbers; the sample used was a marble fragment.

Table 5.1. Analytical lines used [Lazic et al 2004] for calibration on CaCO$_3$ doped powders (pressed), the obtained correlation factor and calibration range for different elements

Element	Line (nm)	Correlation R	Calibration range (ppm)
Ca	300.69	0.997	310000-400000
C	247.86	0.951	100000-120000
Al	309.28	0.992	370-16000
Si	288.16	0.994	1600-69000
Fe	438.35	0.976	160-8100
Ti	498.17	0.994	38-640
Mg	285.21	0.997	150-2100
Mn	403.08	0.999	90-2400
Ba	493.41[i]	0.997	0-210
Cu	327.40	0.999	170-690
Na	589.60	0.958	1000 – 3500

(i) = ionic line

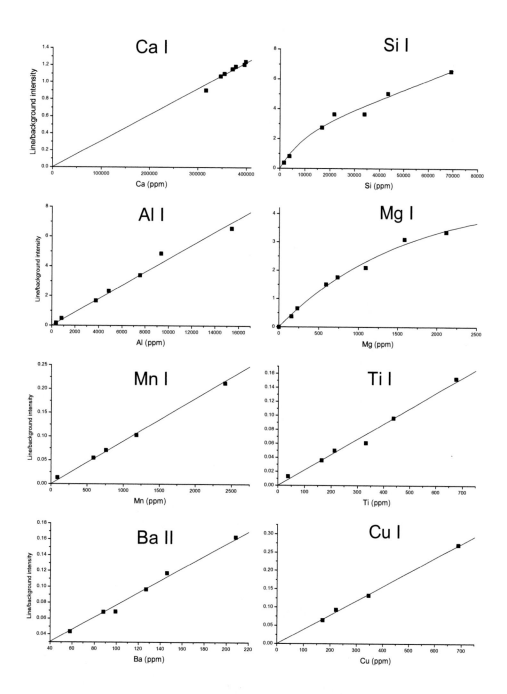

Figure 5.5. Example of calibration plots from doped CaCO3 powders (pressed); line intensity refers to the integrated line intensity.

Calibration for the quantitative LIBS analysis is time consuming and valid only for a certain class of the samples.

Another approach for retrieving the element concentrations by LIBS is the so-called Calibration-Free (CF) procedure [Tognoni et al. 2002, Corsi et al. 2001], which is based on a simultaneous detection of all major elements in the sample and the assumption of plasma in LTE. The plasma temperature T must be first determined from Boltzmann plot (see section 3.2) and then the plasma electron density (section 3.3).

Once the plasma parameters are known, the species concentrations in the plasma is determined from the emission of one neutral (or ionic) line (Equation 3.2) and the number density of the corresponding ions (i.e., atoms) is then calculated from the Saha equation (Equation 3.3). The concentration of the element α in the sample is considered proportional to the sum of atomic and ionic (single ionized) particles:

$$C_\alpha = b_\alpha{'} (N_\alpha^I + N_\alpha^{II})$$ (5.1)

Here, $b_\alpha{'}$ is an experimental constant for one element, which is assumed to be equal for all the elements in the CF normalization:

$$\sum_\alpha C_\alpha = 1$$ (5.2)

The CF procedure requires that all the major sample elements are simultaneously detected by LIBS and was successfully applied for analyses of aluminum and precious alloys [Corsi et al. 2001]. However, there are elements that are difficult to be detected by LIBS even at high concentrations, such as sulfur and chlorine, and there are the elements that suffer from interference from the surrounding atmosphere, such as oxygen for the measurements in air. Whenever the samples contain oxides which form is known for the major matrix elements, oxygen contribution could be calculated [Colao et al. 2002] and inserted in Equation 5.2. CF LIBS applied on unknown samples with a significant content of unidentified or interfering elements, could to a large measurement errors. For some elements in soils, the measured concentration by CF_LIBS is even one order of magnitude different from the real one [Colao et al. 2004b]. Further limitations of CF-LIBS accuracy are due to uncertainty of available databases for less studied elements, which in particular could contribute to analytical errors through the calculation of partition functions, included in eqns. 3.2-3.3. Finally, the CF procedure assumes the preservation of the sample stochiometry in the gas phase, which is often missing for some classes of materials, as for example copper alloys containing the zinc [Fornarini et al. 2005].

Beside the two above methods for obtaining quantitative LIBS results, a mixed approach including both the initial calibration and CF procedure have been also reported [Colao et al. 2002]. Here, the initial calibration was used to retrieve the coefficients b'_α for each analyzed element, which resulted to be different up to a factor three from one element to another. Then the CF normalization was applied following the revised formula:

$$\sum_\alpha C_\alpha / b'_\alpha = 1$$ (5.3)

In this way, the uncertainties in the atomic data bases and matrix effect are partially compensated. The described hybrid method was applied for analyses of the multi-layered renaissance ceramics [Colao et al. 2002c].

5.3. Corrections for Variable Plasma Parameters

In order to analyze quantitatively the samples starting from calibration on reference materials whose LIBS plasmas have different temperature and electron density, correction coefficients for the element concentrations have been derived [Lazic et al. 2004] starting from the assumptions:

- The whole plasma emission is collected by the receiver optics
- Within the LIBS signal acquisition window the second and higher ionization stages in the plasma can be neglected
- Within the chosen acquisition window the plasma is in LTE

If the line intensity in eqn 3.2 is normalized to the continuum plasma emission, which here results proportional to the ablated volume, N_α assumes significance of the species number density in plasma with the temperature T produced by ablation of the unit sample volume.

It had been demonstrated that the concentration of element α in the sample measured through the atomic emission is N_α (with $N_\alpha \equiv N_\alpha^I$ in eqn 3.2), is proportional to the analytical line intensity, where the linear coefficient depends both on T and N_e [Lazic et al 2004]:

$$I_\alpha = b^{ki} \frac{e^{-E_k/kT}/U_\alpha(T)}{1+f_{2\alpha}(T,N_e)} C_\alpha \equiv A_\alpha(T,N_e) \, C_\alpha \qquad (5.4)$$

where b^{ki} is an experimental constant for given transition, $f_2(T,N_e)$ corresponds to the number density ratio of ions and atoms in the plasma (see Equation 3.3).

For the sake of readability index α will be omitted from following formulas.

After generating the calibrations plots $I(C^{REF})$ for single elements by using standards with different concentrations C^{REF}, the calibration coefficients A^{REF} are determined for corresponding T^{REF} and N_e^{REF} i.e. temperature and electron density in the case of reference samples. For an unknown sample having T^X and N_e^X different than in the case of reference materials, the concentration C^{REF} found from the calibration graph should be corrected for the factor F^X that accounts for differences in plasma parameters:

$$C^X = F^X \cdot C^{REF} \qquad (5.5)$$

The correction coefficient F^X for the concentration measurements through an atomic line emission is given by the next formula:

$$F^X = \frac{A^{REF}}{A^X} = \frac{1+f_2(T^X,N_e^X)}{1+f_2(T^{REF},N_e^{REF})} \cdot \frac{U^I(T^X)}{U^I(T^{REF})} \cdot \frac{e^{-E_k/kT^{REF}}}{e^{-E_k/kT^X}} \qquad (5.6)$$

For element concentration measured through an ionic line emission, the correction factor F_i^X is:

$$F_i^X = \frac{1+1/f_2(T^X,N_e^X)}{1+1/f_2(T^{REF},N_e^{REF})} \cdot \frac{U^{II}(T^X)}{U^{II}(T^{REF})} \cdot \frac{e^{-E_k/kT^{REF}}}{e^{-E_k/kT^X}} \qquad (5.7)$$

Although these correction factors are derived neglecting the line self-absorption, this whenever present can be partially compensated by a non-linear calibration curve.

Basing on eqns. (5.6-5.7) the correction coefficients were calculated for each detected element in single marble sample starting from the measured LIBS plasma temperature and electron density. For some elements these coefficients are very different from unity. When the corrections are not applied the LIBS measured concentrations are particularly overestimated for Ca (up to factor 3), Ti and Mn and in some cases significantly underestimated for Ba. The measured LIBS concentrations of the most characteristic marble elements after applying the described corrections are compared with the values obtained by the other techniques (SEM-EDX and ICP-OES) giving the satisfactory results [Lazic et al. 2004].

Table 5.2: Plasma temperature, electron density and calculated correction coefficients for different elements relative to the calibration standards (CaCO₃ powder pellets) and analyzed marble samples

Sample	REF	aks1-b	aks1-c	At13-b	at13-c	Naxos-b	Naxos-c	Paros-b	Paros-c
T(K)	9400	8100	8100	8000	8000	7500	7500	7600	7600
$Ne\cdot10^{17}$ (cm^3)	10.6	4.5	6.5	6.4	6.0	6.6	8.6	5.0	3.6
Correction factors F^X									
Ca E_j=6.113	1.00	0.67	0.55	0.52	0.53	0.38	0.35	0.45	0.52
C E_j=11.26	1.00	1.11	1.12	1.13	1.13	1.16	1.17	1.15	1.15
Si E_j=8.152	1.00	1.10	1.08	1.10	1.10	1.17	1.17	1.16	1.16
Al E_j=5.986	1.00	0.99	0.93	0.93	0.94	0.94	0.92	0.95	0.99
Fe E_j=7.902	1.00	0.86	0.84	0.83	0.83	0.81	0.80	0.82	0.84
Mg E_j=7.646	1.00	0.85	0.80	0.79	0.80	0.78	0.76	0.79	0.82

Mn $E_j=7.437$	1.00	0.61	0.56	0.53	0.54	0.43	0.42	0.46	0.49
Ti $E_j=6.82$	1.00	0.74	0.65	0.63	0.64	0.54	0.52	0.59	0.63
Ba$^{(i)}$ $E_j=5.212$	1.00	1.13	1.24	1.28	1.25	1.62	1.82	1.39	1.26
Cu $E_j=7.726$	1.00	0.98	0.96	0.96	0.96	0.98	0.98	0.98	0.99

(i) = ionic analytical line
suffix (-b) = bulk, (-c) = crust

6. LIBS APPLICATION FOR SOIL ANALYSIS

6.1. General

Characterization of sediments and soils is particularly important in geological studies, in agriculture and for environment monitoring and protection, as for example - determination of pollution by heavy elements. There are a number of well-established analytical techniques that can be applied on soils, such as: X-ray fluorescence, atomic absorption and emission spectroscopy, mass spectrometry and neutron activation analysis. In comparison with these standard techniques, at the current state of development, LIBS can compete only under limited circumstances in terms of detection sensitivity, precision and accuracy. However, the widespread use of LIBS also for soil analyses, can be attributed to its several appealing characteristics, including real-time multi-elemental analysis, in-situ field operation without sample preparation, as well as a possibility to operate remote measurements.

LIBS spectra from the soils/sediments are particularly rich because they contain the emission lines from numerous elements (Figure 6.1). These spectra may easily have more than 1000 lines, most of them at the wavelengths below 550 nm. Due to spectral interferences (Figure 6.2), the number useful analytical lines for some elements are limited. While for the major sample constituents the resonant transitions should possibly be avoided, as they lead to the saturation of the calibration curves, the strong resonant emissions are important for detecting the trace elements.

Main difficulties in quantitative LIBS analysis of soils are related to the matrix-effect (see section 4.1.1) given the great variability of existing soil/sediment types, their granulometry and the other physical properties. An additional problem in soil analysis is caused by variable water content in the natural samples. Generally, the LIBS signal decreases with increasing the water content [Wisbrun et al. 1994] since a large part of the laser energy is wasted for heating and evaporating the water. However, an increase of water content up to a few percent can enhance the LIBS signal [Bublitz et al. 2001]: this was attributed to a formation of the aquatic colloids, which can lower the ablation threshold, and to a reduction of dust particle density above the surface. These particles are removed and dispersed by the laser-induced shock waves and act as the radiation scatterers. The influence of water content on the LIBS signal must be considered non-linear and responsible for the changes in the plasma properties (shape, temperature and electron density) and in the aerosol formation.

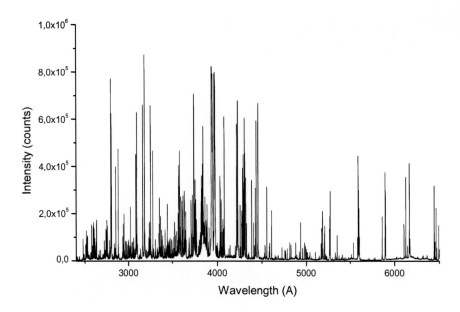

Figure 6.1. Example of the LIBS spectra from a soil sample.

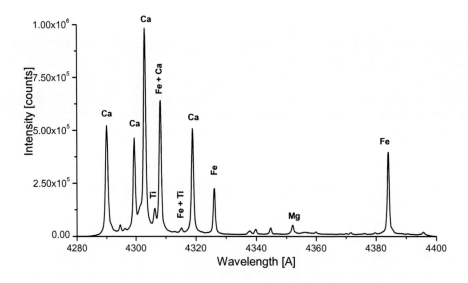

Figure 6.2. Portion of LIBS spectra from a soil sample.

Consequently, when analyzing a wide span of soil types, very different signal intensities could be observed under the same experimental conditions. One approach for quantitative LIBS analyses on an unknown soil is to include into calibration very different types of the soil standards [Lazic et al. 2001] and to apply some kind of normalization (section 5), which partially compensates for variable ablation rates.

In addition, if the calibration curves include both ionic and atomic species in the plasma, one of them could be calculated from Saha equation after determining the plasma parameters, an acceptable accuracy could be achieved on a wide range of soil matrices. In this way, an analytical accuracy better than 20% was obtained for most of the elements in several soil samples [Lazic et al. 2001]. Although this accuracy is considered too low for the laboratory analyses, it can be satisfactory for in-field measurements, as for example during fast surface mapping of the contaminants. Once the polluted areas are identified by LIBS, the samples can be taken for more accurate laboratory measurements also by other, well established analytical techniques. For the measurements of relative element concentration changes in special cases (for example during surface mapping in a delimited area), it is usually sufficient to observe only a variation of the line intensity ratios, without performing a laborious laboratory calibration for each of the element of interest.

In laboratories, often the soil/sediment materials are first pressed into pellets and then analyzed by LIBS. In this way, the matrix effect due to very different surface compactness of the natural samples is reduced, as well as the uncertainties due to uncontrolled surface roughness. Commonly, the soil pellet is rotated in order to sample always a fresh area and to minimize the negative effects of the aerosol formation and material re-deposition (section 5).

As discussed earlier, a higher LIBS accuracy can be obtained if the calibration and successive sample analyses are performed on matrix similar samples, as for example only clays or only organic rich soils. Standard reference soil/sediment samples are supplied in powders, previously machined, passed though a sieve and dried. For these standards only a limited number of elements are certified. The standard soil materials are often doped in laboratory in order to increase the number of reference samples for the calibration, and in particular, to extend the calibration range for some certified elements.

On the other hand, the Calibration-Free method (section 5) has been demonstrated to be unsuccessful for quantitative soil analyses as so determined concentrations for some elements have values even for one order of magnitude different from the real ones [Colao et al. 2004b].

In following, the examples of LIBS applications to soil/sediment analyses are reported, including the experimental conditions, analytical procedures and final results.

6.2. Quantitative Analyses of Soils with Unknown Matrices

One of the LIBS methods for quantitative analysis of soils with different origins is proposed in [Lazic et al. 2001] and discussed in following. This method includes the usual plasma modeling at LTE, and calculations of average plasma temperature and electron density. The spectral normalization was applied in order to reduce the effects related to the variable ablation rates, expected for the soils of different nature and for rough sample surfaces in natural state. The computational algorithm takes into account only atomic species and their first ionization states, which is sufficient for the plasma temperatures measured in the experiments (7500-8500 K). The calibration curves were generated for the each element of interest, by using the certified samples with different provenience and matrix composition. In this paper, also a simple model has been developed, which takes into account line re-absorption and radiative contributions from space regions with different plasma densities. Its application permits us to obtain a simple, non-linear relation between the LIBS measured concentration and the certified values. The experimentally obtained fitting coefficients are specific for a given experimental layout and for the spectroscopic data base used, and they were successively applied in the analytical LIBS measurements of unknown soil matrices. This LIBS method was then tested on a priori unknown samples, and in most of the case gave uncertainties varying from 15 to 20% over a large concentration range, covering several orders of magnitude for some elements. For strong analytical lines, as for example the resonant lines form Al and Na, the measuring error was high up to 40%. The error bar depends on element type, on the concentration value in the sample and also on the number of certified samples included in the initial calibration. Here presented results are already significant for some field application where a significant matrix variation from one sample to another is common.

6.2.1. Experimental

The plasma was produced by an Nd:YAG laser operated at III harmonic (355 nm), with pulse duration of 8 ns and repetition rate of 10 Hz. The laser beam was focused perpendicularly onto the sample surfaced by a plano-convex lens with focal length of 150 mm. The light emitted by the plasma was collected by wide angle receiver optics, mounted at 15° with respect to the laser beam axis. The collected signal was carried by an optical-fiber bundle to the entrance slit (equivalent slit width 0.1 mm) of a 550 mm monochromator, used with the grating containing 2400 grooves/mm. At the exit plane, an ICCD records the LIBS spectra. All the equipment was controlled by custom written software routines, working in a Windows environment.

Initially, test measurements were performed by varying the laser energy between 6-10 mJ. By applying the analytical procedure described in following, the quantitative measurements resulted independent on the laser energy in this range. Then, the middle energy value (8 mJ) was chosen for the further experimentations. The spectra were acquired accumulating the signal over 10 or 20 laser pulses. The acquisition delay with respect to the laser pulse was 300 ns and the gate width was 1000 ns. Under these conditions it was possible to observe the produced plasma at LTE.

The investigated samples were pressed into pellets and inserted on a rotating holder in order to expose a regular surface, thus to compensate partially the local sample inhomogeneities, to minimize crater development and effects of the aerosol formation. The sample rotating speed, number of laser shots and other experimental parameters were adjusted to optimize the measurement reproducibility [Wisbrun et al. 1994].

The reference samples used for LIBS calibration, having representative contents of heavy metals, were supplied by different International Institutes for Reference Materials. As summarized in Table 6.1, these samples have very different natures and provenience, even a sample of urban particulate was included.

For validation of the LIBS data analysis procedure as a quantitative technique, a sample of Antarctic sediment collected during the XI Campaign and certified by MURST was examined. This sample, further called the Antarctic reference sediment, was not included in the set used for the initial calibration. LIBS quantitative analyses were further applied to unknown samples extracted from a 25- cm-long core of Antarctic sediments collected during the XIII Campaign. The core, extracted from the Ross Sea (Latitude 72°10'50", longitude 172°47'77", depth 512 m) was divided in seven parts sampled at depth layers, dried and sifted through a net with 0.2 mm diameter holes.

Table 6.1: Certified samples used for the calibration

Description	Origin	Number of samples
Urban particulate matter	St Louis, Missouri	1
Marine sediment	Gulf of St. Lawrence	1
Marine sediment	Harbour of Esquimalt	1
Estuarine sediment	Chesapeake Bay	1
Marine deposit	Sicilia (I)	1
Glacial sediment	Galles (UK)	1
Loess	Normandia /F)	2
Sludge from city water treatment	-	3

6.2.2. Data Analyses Procedure

Each fraction of acquired spectra was firstly corrected automatically in wavelength and then in intensity. The last correction was referred to the overall instrument response as a function of wavelength, previously measured by using a lamp with certified spectral irradiance. Corrected portions of spectra were then assembled to cover the spectral range from 240 to 650 nm. Integral line intensities were calculated from the peaks fitted by pseudo Voigt distribution. In order to reduce the influence of the sample properties, of the laser energy oscillations and of pulse number on quantitative analyses, the retrieved line intensities were normalized by integrated plasma emission intensity, automatically calculated from the full spectra. So obtained line intensities, dependent on the species density in the plasma (Equation 3.2), were also divided by a fixed constant which gives a correction factor b (Equation 5.1) close to the unit in the case of Fe.

The temperature calculation was performed for each spectrum on weakly absorbing Fe lines free of overlap, and then inserted into Equation (3.2) in order to estimate both Fe I and Fe II content in the plasma from their emission lines. Their concentration ratio allowed retrieving the electron density according to the Saha equation (3.3). The measured plasma temperature for different samples was between 7500 and 8500 K, while the estimated electron density was in the order of 10^{16} cm^3.

Once the plasma temperature and electron density are known, the concentration of other neutral species is determined from the integral line intensities and the program automatically performs a calculation of first ionization state and of the total element concentration. The concentrations of the atoms were obtained using the spectral lines free of overlap present in the NIST database [NIST] and completed by the atomic data from [Kurucz].

After initial generation of calibration curves, their fittings were performed by a non-linear equation:

$$C_m = \frac{a_1}{a_2}\left(1 - e^{-a_2 C_E}\right) + a_3 C_E \qquad (6.1)$$

Where C_m is the LIBS raw measured element concentration, obtained from Equation (5.1), while C_E is the effective element concentration in the sample (for standards, it corresponds to the certified values).

The Equation (6.1) was derived assuming that the observed plasma volume is not homogeneous, but containing a hot core and the external, cooler layer. The later is responsible for the re-absorption of the line transitions terminating to low energy levels. The first term of Equation (6.1) corresponds to the optically thick plasma and the second, to the volume non-absorbing at the wavelength of the chosen analytical line. If the considered atomic line is weakly absorbing, the coefficient a_3 relative to the optically thin plasma approaches the coefficient b_α from Equation (5.1). The latter should be identical for all the elements under fixed experimental conditions if the used values from the databases are accurate both for the atomic and the first ionization level, the species have the same spatial and temporal distribution and the sample stochiometry is preserved in the plasma.

In the successive analysis of the unknown samples, an element concentration C_m was determined as previously described and then the corresponding concentration in the sample (C_E) was calculated by applying the previously obtained calibration coefficients in Equation 6.

6.2.3. Results – Calibration

Spectra normalization by overall plasma emission significantly reduced the matrix influence on quantitative analyses and this can be observed from the reduced scattering on the calibration graphs (Figure 5.3). Examples of the obtained calibration graphs (the element concentrations expressed as a percentage of the weight) and the corresponding data fitting by Equation 6.1 are given in Figure 6.3. The residual scattering of experimental data along the fitted curves can be attributed both to the residual influence of sample properties and to the errors in computing the line integrals, plasma temperature and electron density.

For most of the elements analyzed the LIBS measured concentration grows non-linearly with the certified concentration values, thus indicating a presence of self-absorption. The non-linearity is particularly pronounced for Ca, Na, Ni, Cu and Ba as their relatively strong lines were involved in the concentration measurements. For Ni (343.36 nm), Cu (327.39) and Ba (553.55), the resonant lines were chosen for the analysis since they are the most intense and allow for obtaining the lowest detection thresholds. For the selected Ba line, the reliable quantitative analyses can be performed only for the concentrations below 0.04%, since above this value the calibration graph is almost completely saturated. Consequently, for higher concentrations of this element it would appropriate to select some less self-absorbed analytical line. Regarding Na, the only line sufficiently strong for identification and free of the spectral overlap is the resonant doublet around 589 nm. Beside the nonlinearity of the calibration graphs, the self absorption of a line can be observed from the line broadening with the concentration. The broadening of the resonant copper line due to the self-absorption is illustrated in Figure 6.4. The line width obtained for the highest elemental concentration is 0.070 nm, which is about 13% larger than one measured at the lowest concentration (0.059 nm). A departure from the Gaussian line fit could be also noticed at increasing elemental concentration, thus indicating a necessity to use Voigt or pseudo-Voigt fit.

We finally mention that the calibration for Fe was checked on different lines included in the Boltzmann plot and they resulted linear, so the temperature measurements were not affected by the self absorption. Similarly, the calibration linearity was checked for Fe II lines used for the electron density measurements through the Equation (3.3).

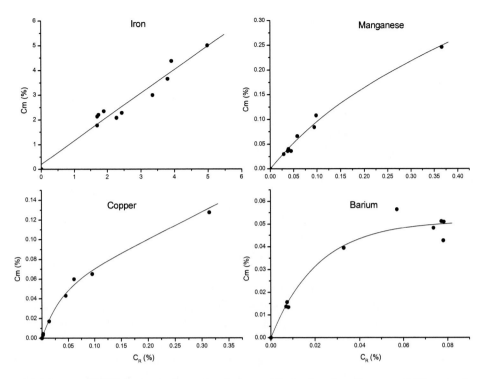

Figure 6.3. Measured LIBS concentration towards the certified values in reference soil-like samples with different matrices.

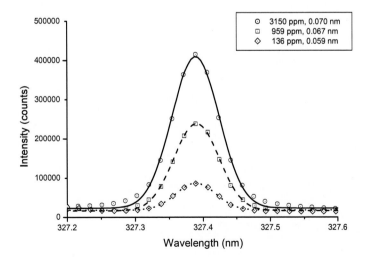

Figure 6.4. Shape and width (FWHM) of Cu line at 327.39 nm measured for three element concentrations.

6.2.4. Results –Quantitative Analysis

For the main elements determined in the Antarctic sediment examined, differences between the measured and certified concentration values, expressed as a percentage of the latter, are shown in Figure 6.5. The measurement error is below 20% for most of the elements analyzed, only for sodium, aluminum and zinc, the obtained concentration errors are above 30%. The LIBS measured concentrations of Al and Na are much lower than in the certified values, which were high and falling in the saturated parts of the calibration graphs. Consequently, large measurement errors can be here expected. In the case of zinc, a large measuring error in excess is due to the partial overlapping of the chosen emission line with the non-weak spectral lines from other elements.

Beside the elements shown in Figure 6.5, also Si and Ti were determined by LIBS, while Cu presence (5.9 ppm) was below the detection threshold. The sum of the measured elements corresponds to the concentration of about 47%. The main remaining undetermined constituent is oxygen, beside other elements not analyzed elements, e.g. C, K, P, S, etc.. By extending the detection range towards infrared, it is possible to measure also K emission through its strong lines at 766.5 nm and 770.0 nm. This element also shows two emission lines around 404 nm, but they are close to one Fe line, which in the case of the examined soils is sufficiently intense to mask K lines by its broad tail. Phosphor has few sufficiently strong lines around 253 nm, again overlapping with Fe emission, so this element can not be easily measured in this spectral region, at least at low concentrations with respect to Fe. On the other hand, Sulfur is a highly reactive element in air and its detection by LIBS require a low pressure surrounding or a prevalence of inert gasses such as for example argon, helium etc.

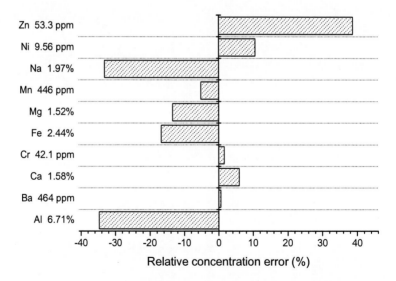

Figure 6.5. Relative concentration errors measured by LIBS on one certified sample.

6.3. Sediment Analysis

Here we present the application of LIBS on board the R.V Italica during the XVI Antarctic campaign (2000–2001), aimed to carry out elemental chemical analysis of marine sediments collected by using different sampling systems. To this end, a compact system has been built, which was suitable to operate also in the presence of mechanical vibrations, induced by the ship motion. It has been demonstrated that a LIBS apparatus, used with a proper analyzing procedure, can be employed for in-situ sediment analyses even in spite of hostile environment characterized by strong vibration instabilities, as in the case of the cruise across the Antarctic Sea. The implementation of LIBS technique for fast screening of the element distribution inside a sediment core, is extremely useful for rapid overviews during local and/or extended monitoring in the frame of the oceanographic campaigns. This allows to tag important sites for more detailed biochemical analyses and to supply further information for biological and geochemical studies. To this respect significant data related to the paleo-climate of the examined sites can be deduced from LIBS analysis. Qualitative and quantitative analyses were performed on dried samples, without any further pre-treatment.

6.3.1. Experimental

The experimental apparatus used during the Antarctic campaign is a more compact version of a measurement facility already described in the previous section (6.2). The whole system was mounted on a support made of anti-vibration material, in order to reduce effects of continuous or accidental vibrations related to the ship's movement. The third harmonic Nd:YAG pulsed laser beam at 355 nm was deflected towards the sample by a dielectric coated mirror, and focused by a 250-mm plano-convex lens on the sample surface. The light emitted by the plasma is collected by suitable receiver optics, mounted above the deflecting mirror, i.e. at 0° with respect to the laser beam axis. axis. A co-axial optical lay-out with wide angle receiving optics was chosen in order to capture emission from entire plasma and so reduce the effects of sample surface irregularities, expected in the case of direct measurements on the sediments in box corer. The collected signal is transferred through an optical-fiber bundle to the entrance slit of a 550 mm monochromator, used with the grating 2400 grooves/mm. At the monochromator exit plane, a gated ICCD records the LIBS spectra. The acquisition delay with respect to the laser pulse was fixed to 300 ns and the gate width was 1000 ns.

The spectra were acquired by averaging the signals generated by 50 laser pulses with energy of approximately 25 mJ. In order to achieve a full online analysis, the measurements of the element distribution inside a core were performed acquiring the spectra only at the limited number of the monochromator central wavelengths, selected since they contain a suitable number of emission lines from the elements of interest. On the other hand, in quantitative analyses of dried and pressed samples, the complete spectra were acquired in the whole range between 240 and 650 nm.

During spectra acquisition, the sample was moved manually in the case of core analyses, or automatically rotated in the case of measurements on the pressed samples. In order to reduce the measuring error and effects of sample inhomogeneities and roughness, the sediment core was scanned by moving the slice from the top to the bottom layers, and then backwards. Sediment samples were collected by a Box Corer or a Gravity Corer. In the first case, the instrument extracts a cylindrical section of sediment, 30 cm diameter and 50 cm high, from the seabed. From this core, a longitudinal slice is sampled for LIBS through a semi-cylindrical box with diameter of 2 cm, in order to obtain an undisturbed sediment depth profile. On this sediment slice, a drying cycle was applied inside an oven at 42°C for at least 48 h. Sample drying was necessary since otherwise the LIBS emission should be dampened by the laser energy loss on the water evaporation competing process. After drying, the element distribution inside the core was directly analyzed by the LIBS technique, by moving manually the slice during measurements. An example of dried sediment slice directly analyzed by LIBS is shown in Figure 6.6. After these measurements, some representative sediment layers were also extracted (approx. 2 g in weight) and pressed at 38 MPa into pellets of 13 mm diameter, to be used for quantitative LIBS analyses.

The sediments extracted by the gravity-corer were not open during the campaign, so it was possible to analyze only a small portion of sample remaining on the top of the corer, which corresponds to the highest sediment depth reached by the corer. Geographic data relevant to the samples extracted by box-corer (BC) and gravity-corer (CAR), are given in Table 6.2 and the map with mooring locations is shown in Figure 6.7.

Figure 6.6. The core of BC3 sample after drying treatment.

Table 6.2: Geographic data relevant to the samples extracted by box-corer (BC) and gravity-corer (CAR)

Sample	Date and time UTC	Position	Depth [m]	Note
BC1	8 Jan 2001-19.24h	61° 35'S–174°28'E	4153	
BC2	9 Jan 2001-23:11h	64° 30'S–176°50'E	3444	
BC3	17 Jan 2001-20.12h	75° 10'S–164° 23'E	1213	Mooring D1
BC4	20 Jan 2001-20:26h	73° 52'S–174° 44'E	530	Mooring B
BC5	28 Jan 2001-9:15h	71° 14'S–170° 26'E	605	
CAR6	28 Jan 2001-10:51h	72° 15'S–170° 44'E	409	
CAR7	29 Jan 2001-1:17h	74° 59'S–171° 50'E	550	
CAR8	29 Jan 2001-6:26h	75° 44'S–171° 25'E	571	
BC9	30 Jan 2001-4:42h	76° 42'S–168° 48'E	835	Mooring A
CAR10	31 Jan 2001-1:46h	77° 05'S–166° 57'E	868	

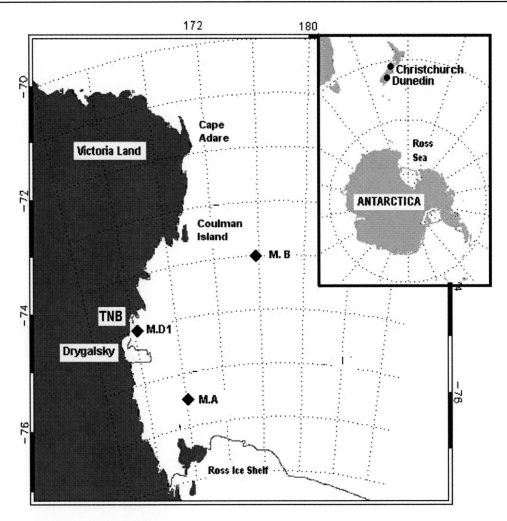

Figure 6.7. Position of the mooring sites where the cores used in element distribution analyses were extracted.

6.3.2 Element Distribution inside the Sediment Core

Before quantitative analyses of the sediment samples, a fast semi-quantitative scanning of the element distribution inside the core was carried out. These measurements allow for rapidly identifying the most interesting sites and samples for further detailed examination, and give straightforward information to geologists and biologists interested in the elemental stratification along the sampled column.

The line intensities used both for quantitative and element distribution measurements are referred to the integral of line intensity fitted by a pseudo-Voigt profile. The latter was chosen instead of Voigt profile due to the simpler data processing, which makes possible fast fitting of all the emission lines in the acquired spectra.

Instead of simply comparing the line intensities for single sediment layers, the distribution analyses were based on calculation of the species concentration in the layer through Equation 3.2 after assuming the same experimental constant for all the elements and a fixed plasma temperature. The latter parameter was determined at the central monochromator wavelength of 388 nm, where different atomic Fe emission lines are present and by which the plasma temperature was obtained through the Boltzmann plot. In all the measurements on the sediment cores the calculated temperature was in the range of 7700±400 K, so the plasma temperature could be considered constant within the measuring error. Knowing the plasma temperature, it was possible to compare the element concentrations corresponding to different emission lines of the same element, if they are present in the same spectral region, and to reduce errors due to the line fitting procedure. Wherever more than two not overlapping elemental lines were present, as in the case of Fe and Ti, the analyzing software eliminates automatically those for which the measured concentration results in a scattering of more than 30% with respect to the mean value and the final average value is considered as a result of the measurement. The lines used for distribution measurements of different elements are reported in Table 6.3.

Table 6.3: Wavelength of the emission lines (nm) used for the element distribution analyses

Element	Central monochromator wavelength (nm)		
	388	**398**	**557**
Fe	385.00		
	385.64		
	385.99		
	387.25		
	388.65		
	389.97		
Al		394.39	
		396.18	
Si	390.60		
Mg	383.82		
	383.22		
Ca II		396.82	
		393.34	
Ti		398.18	
		398.93	
		399.82	
		402.46	
Mn		403.06	
		403.30	
Ba			553.57

In order to reduce the matrix effects, surface roughness and laser instabilities on the final results, intensities of the emission lines used for retrieving concentrations were firstly normalized to the intensity of the plasma continuum averaged over the acquired spectral range. In Figure 6.8 there is an example of two spectra acquired at same corresponding sediment depth by scanning the core forwards and backwards, respectively. It could be noticed that although the absolute line intensities strongly differ in the two case because of missing surface planarity. However, after the line normalization the measured element concentrations (in this case of Mg, Si and Fe) remain the same within an error of 15%. The concentrations obtained in this way do not correspond to real absolute values for which the previously described analyzing procedure (section 5) requires the acquisition throughout the whole spectral range between 240 and 650 nm, and the application of calibration coefficients. Nevertheless, these relative concentration values correctly describe the variations of the element distribution along the sediment column as far as the line self-absorption effects could be neglected and the atomic-to-ionic ratio in plasma does not change significantly. The element concentration at a single core depth was then calculated as a mean value over scanning the core forwards and backwards, while the error bar was calculated as dispersion around the mean value.

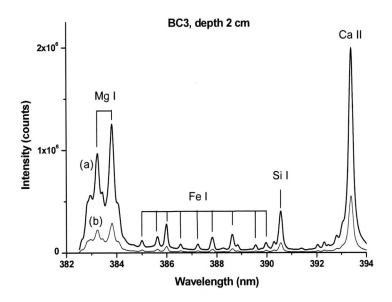

Figure 6.8. Spectral portion acquired on sample BC3 at depth of 2 cm: (a) forward scanning; and (b) backward scanning.

The method described here for the calculation of element distribution has the following advantages: independence from laser energy fluctuations; uncritical optimal focusing at the irregular sample surface; strong reduction of the effects of different water contents, as well as of other sample physical properties; reduction of errors due to the mathematical fitting of emission lines for elements with more than two lines free of overlap; no influence of the measuring errors due to spectral calibration of the entire system; and rapid implementation by an automatic data analysis software. The main disadvantages of the described method rely on the impossibility to obtain absolute concentration values or the concentration ratio for different elements. Although the electron density should not vary significantly from one measurement to another, also because the temperature is approximately constant and the samples are similar, its oscillation is a further source of error due to changes of the atom-to-ion ratio. Furthermore, in the case of self-absorption, which could occur for the used Ca II, Al

I and Ba I lines, the real distribution changes could be more pronounced than the measured ones, but a correction algorithm cannot be settled without a complete quantitative analysis (section 6.2).

In Figure 6.9 the measured element distributions are shown for core BC3. The values reported on the graph are normalized to the mean element concentration value over the whole sediment core. The core BC3 extracted from Ross Sea, close to the coast of Baia Terra Nova, shows a rather uniform distribution of Fe, Al, Mg and Ti. On the other hand, Si and Ba, which can be mainly correlated to bio-productivity [De Mater 1981, Dehairs et al. 1991], are clearly decreasing in the uppermost 10–12 cm. The higher content of Si in the upper layers may be related to a higher bio-productivity (diatoms, radiolarians and sponge spicules) and perhaps to the diversification of foraminifers during Holocene [Dehairs et al. 1991], with the appearance of arenaceous forms. The measured Ba is reduced even by 70% from this top sediment layer, after which it shows some characteristic structures with depth. The minor contribution of sedimentary barium derives from the lithogenic mineral fraction, usual feldspars, while the main contribution comes from barite precipitation due to the settling of diatoms through the water column [Dehairs et al. 1991]. To this respect, it is worth mentioning that the samples come from an area characterized by phytoplankton blooms in recent periods [Saggiomo et al. 2000]. Mn, which is a tracer of redox conditions and whose distribution can be used to infer palaeo-productivity [Kumar et al. 1995], after the peak close to the surface shows another maximum at approximately 25 cm depth, corresponding to a darker color of the sediment. Since the core was extracted very close to the continent, the element distribution could be strongly influenced by supply of terrigeneous material. Aluminum, the most characteristic element of alumino-silicates, is assumed to represent the terrigeneous portion of clay minerals [Collier et al. 2000], whose variation inside the sediments could be attributed to glacial–interglacial cycles with alternate rise and fall in the supply of continental materials.

To estimate barium originated from the precipitation of diatoms, due to the missing absolute concentration values the normalization to Al had to be applied instead of the barium excess calculation. Both Ba and Mn, when normalized to Al concentration (Figure 6.10), show a similar behavior with sediment depth. Recent studies from the Southern Ocean have shown that south of the Antarctic Polar Front, the glacial paleo-productivity and consequently, the sedimentation rate was lower than during the interglacial period. With the hypothesis that the first common maximum in Figure 6.10 could be ascribed to the Holocene warmth peak (Hypsithermal) approximately 6 ky before present (BP), then the corresponding average sedimentation rate turns out to be 3.7 cm/ky, in agreement with the values between 2.2 and 3.8 cm/ky already published for some areas of the Ross Sea [Ceccaroni et al. 1998, Brambati et al. 2000], regarding Holocene period. Assuming this sedimentation rate as a constant, the minima in the Ba distribution, observed at a depth of 38 cm, would correspond to 10.3 ky BP and may be attributed to the cooling episode which occurred at approximately 11 ky BP.

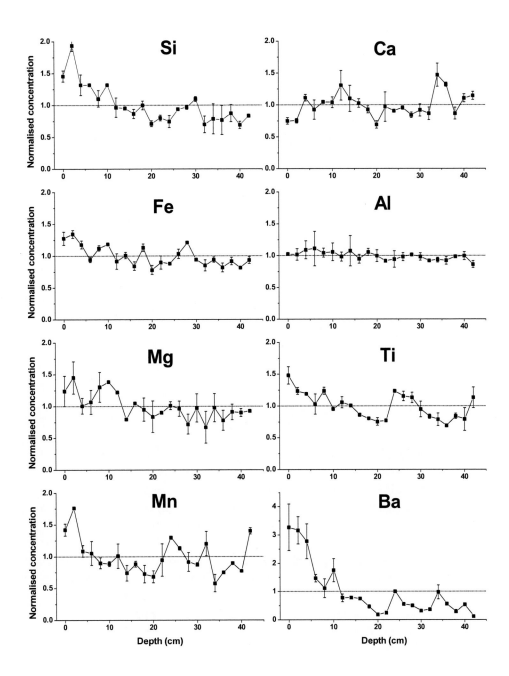

Figure 6.9. LIBS measured relative element distribution inside the sediment core BC3.

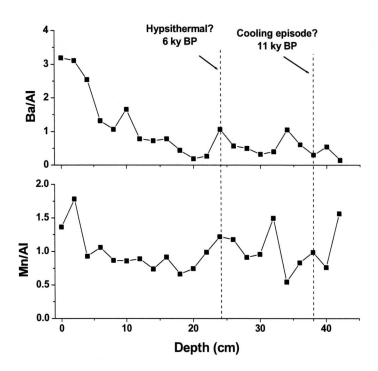

Figure 6.10. Measured distribution of Ba and Mn normalized on relative aluminum content, for BC3 core.

Other extracted sediment cores were analyzed on the same way. As for example, the core BC4, extracted far away from the coast (Mooring B), show the less pronounced decrease of Si and Ba with depth. All other elements, except Ca, are slowly changing, with some oscillations and a decreasing tendency with depth. Calcium-rich deeper layers were observed from the spectra and this was attributed to the ancient presence of organisms characterized by a calcium carbonate skeleton such as calcareous foraminifer algae, mollusks, corals, etc., found in the region [Kumar et al. 1995]. Both elements Ba and Mn, once normalized to Al concentration, show characteristic minima at a depth of approximately 32 cm. Assuming that the minima is associated to the cooling episode, the corresponding average sedimentation rate is 2.9 cm/ky. At this accumulation rate, the Holocene Hypsithermal would correspond to a depth of 18 cm, where the element distributions normalized on Al show slightly higher values. At this depth also Fe has a maximum that may be related to the abundance of nutrients in water.

The site of Mooring A, where the core BC9 was extracted, is located much more southerly than the other investigated sites, and very close to the Ross Ice Shelf. Here, an alternation of Si and Ca was observed in the sediment layers, indicating successive interglacial periods (with a warmer climate and a partial melt of the Ice Shelf) when different microorganisms dominated. The LIBS estimated sedimentation rate is rate is 3.1 cm/ky during Holocene.

6.3.3. Quantitative Sediment Analyses

The adopted procedure for quantitative LIBS analyses was already described in the section 5.2. and the calibration coefficients therein derived were used also here. The results obtained on top layers from two sediment cores (BC4 and BC9) are compared to the element content of one reference sediment sample extracted from the Ross Sea at a depth of 80 m, during the XI Italian Antarctic campaign, and certified by the Italian Ministry of University and Research (MURST). Generally, the measured element quantities are comparable to those of the reference sample except for significant differences in Al concentration.

Successively, the extracted sediment slices were divided into layers, pressed into pellets and analyzed in laboratory. In this way, the absolute distributions of the elements given in Table 6.4 were also retrieved.

Table 6.4: Element concentration [%wt] obtained by quantitative LIBS analyses, compared to the reference sample of the Antarctic sediment extracted from different site and in different period

Element	Reference	BC4	BC9
Si	33.5	28.0	35.6
Fe	2.44	1.82	1.32
Ca	1.78	0.61	0.70
Al	6.71	1.26	1.92
Mg	1.52	0.81	1.40
Mn	0.0446	0.016	0.0082
Ti	0.029	0.034	0.0055
Ba	0.051	0.010	0.034
Cr	0.0042	0.0052	0.0037
Ni	0.00096	-	-
Cu	0.00058	0.0019	0.0021

6.4 Other Examples

6.4.1. Portable LIBS System for Detection of Soil Contamination by Lead

Development of here described instrument [Wainner et al 2001] was sponsored by U.S. Army Research office. The system was optimized for measuring Pb concentration in soils and paints, basing on the detection of strong Pb line at 405.8 nm.

The whole system, containing the components listed in following, is contained in 23x51x38 cm aluminum case and powered by a standard 12 V snow-mobile battery:

- Passively Q-Switched laser at 1064 nm, pulse energy and width are 15 mJ and 4 ns respectively
- Focusing lens for the laser radiation
- Fiber bundle for direct collecting of the plasma emission (without lenses), with round diameter 0.5 mm and the individual fibers arranges into an array at the exit, to simulate 0.1 mm wide spectrometer slit

- A single compact spectrometer unit was employed, containing a thermoelectrically cooled CCD. The system resolution is relatively low – 0.35 nm covering a 20 nm wide spectral range.
- The optical head is inserted into a case unit, which terminates in a flat plate with a hole centered at the focal laser point. During the measurements, this plate is in a direct contact with the ground (soil).

The chosen Pb line is close to some emission lines from Ti, which is normally present in soils and interferes in the low-resolution spectra. However, these Ti lines are generally weak in the soil samples, and the closest normally detectable line belong to Fe at 406.4 nm, sufficiently far away to be resolved from the Pb emission. The obtained detection threshold with the portable system applied on the soils was less than 0.01% (not specified). This value is well above the legal limits, but the system was used to map a highly contaminated area close to one dismantled military site, which contained a furnace and where Pb surface concentration varied up to 18%. The area mapped by the LIBS system, although semi-quantitatively, allowed to determine a downwind direction where a higher amount of the furnace emission was deposited. In addition, it made possible to identify a relatively distant and contaminated site from the furnace, probably used for the ash disposal.

6.4.2. Determination of Total Carbon and Nitrogen in Soils

The characterization of Soil Organic Matter (SOM) is important as this matter is a nutrient source and a substrate for microbial activity. Beside the importance of measuring SOM due to its influence on the plant growth, there is also a necessity to monitor the carbon content in soils and plants in order to evaluate processes of atmospheric CO_2 (greenhouse gas) sinking by them and the possibilities to make this sinking more efficient.

In the example here discussed [Martin et al. 2003], the LIBS was applied for measuring carbon and nitrogen presence in the soils. The spectra were produced by an Nd:YAG laser operated at 266 nm, with pulse energy of 23 mJ. A high resolution scanning monochromator was used, equipped with an ICCD. The soil samples of the same type and with known carbon content, were pressed into pellets and moved during analysis in order to sample always fresh points. The spectra were averaged over 10 laser shots.

Carbon content was measured by LIBS through its atomic line at 247.9 nm. This line is a relatively close to one Fe I line (at 248.4 nm), and as iron is always present in the soils it is important to record the high resolution spectra in order to distinguish well these lines. The calibration for quantitative carbon analysis was performed by simple plotting the line integral intensity versus the known concentration. A matrix effect is not expected to be pronounced on the analyzed samples, and the points on the calibration graph are well correlated. For the measurements on different soil types, better correlation was obtained when normalizing C line on one of the detected Si lines.

Nitrogen content was measured by the N II lines around 500 nm. At low laser energies as here used, N II emission did not appear on other type of the samples free of this element, so the influence of the nitrogen coming from the surrounding air was excluded.

After acid washing of the samples, aimed to remove the inorganic carbon component, the LIBS measured nitrogen concentration did not change as it is mainly bonded to the organic compounds. Then repeating the measurements of carbon allowed to determine its concentration due to presence of the organic compounds.

This work points out the possibility to employ LIBS, also as a field instrument, for environmental monitoring of soil carbon and nitrogen.

6.4.3. Remote 3-D Mapping of Environmental Interest

LIBS measurements can also be applied for spatial analyses that include both surface and sub-surface mapping. Example here discussed regards remote LIBS sensing, in coastal scenario subjected to a high industrial activity [Lopez-Moreno 2004]. The samples analyzed included soils, rocks and vegetation.

The LIBS system used for the remote measurements employs a Q-Switched Nd:YAG laser operated in IR and with energy of 1500 mJ per pulse. The optical system, containing a telescope, has a co-axial geometry. The collected plasma emission is resolved by a high resolution spectrometer, equipped with an ICCD.

Surface scanning of the samples was performed across the area large up to 20x18 mm. At a single surface point a spectrum was recorded after each laser shot in order to obtain the element distribution in depth.

In the case of remote analysis of one dock rock, the element mapping was performed for depths up to 0.54 mm. A lateral and in-depth estimated resolution was about 1.5 mm and 0.01 mm respectively. A 3-D image of the sample composition was reconstructed for different elements. In the case of Fe and Cr, the element concentration was retrieved after an opportune calibration. At a stand-off distance of 12 m from the spectrometer to the sample, limits of detection in the order of 0.2% have been obtained for both elements. For the other elements considered, the emission intensities of the characteristic lines were reproduced in 3-D. It was found that both Cr and Fe decreases with the depth on a similar way, thus indicating the external contamination. The maximum measured surface concentrations of these elements were about 1.8% for Cr and 7% for Fe. By mapping contemporary Ca presence (the main rock constituents), it was possible to determine also a depth of penetration of these contaminants (Fe and Cr) inside the bulk material.

From these measurements, it was also possible to associate the concentration measured to the sample orientation and to the surrounding environment. The analogue measurements on other sample types evidenced that the analytical results are influenced by moisture and salinity, and that the pollutants have the high diffusivity inside porous, soil materials.

6.4.4. Combined LIBS-LIF Measurements for Analysis of Heavy Metals in Soils

In order to apply LIBS for monitoring the soil contamination with respect to the regulatory demands, the detection limits (LOD's) for various elements must be below the maximum values that are allowed for unpolluted soils. For most of the heavy metals the LIBS sensibility is sufficient for such kind of analyses, however, there are some elements such as Cd, Hg and Tl which allowed concentrations are lower that the typical LIBS detection thresholds for the same.

In order to include also these elements in the in-situ soil monitoring, a combined use of LIBS-LIF has been proposed [Hilbk-Kortenbruck et al. 2001]. Here, the LIBS signal was generated by a Q-Switched Nd:YAG laser at 1064 nm, which energy was set to 80 mJ after the initial experimental optimization. The LIBS signal was brought by a fiber to a Paschen-Runge spectrometer equipped with 29 photomultipliers for simultaneous detection of the same number of lines. The estimated spectral band-width of each photomultiplier was 0.02 nm. The signals from the photomultipliers were processed by a rapid, gateable multichannel integrator. The optimal gate delay from the laser pulse and gate width was found experimentally, and corresponds to 1 µs and 20 µs respectively.

Another Nd:YAG laser operated at 532 nm was used to pump a dye laser, which radiation was frequency doubled and tuned to the transitions of Tl at 276.8 nm. For the resonant excitation of Cd transition at 228.8 nm, the dye-laser (frequency doubled) emission at 291.5 nm was converted by sum frequency mixing with radiation at 1064 nm emitted from the pumping Nd:YAG laser. The chosen upward transitions for Tl and Cd involve the ground state. The corresponding fluorescent emissions were detected at 351.93 for Tl and 228.8 nm for Cd (the same as the excitation wavelength). LIF excitation was perpendicular to the sample surface, intersecting the LIBS plasma. The optimal delay for LIF detection, with respect to the laser pulse used for the LIBS, was longer than 100 µs and this was attributed to the fact that the maximum number of the heavy atoms must be relaxed to the ground state and that the expanded plasma volume well matches the volume probed by LIF.

The calibration both for LIBS and LIF measurements were obtained from the certified soils, pressed into pellets before the measurements. In Table 6.5 the LOD's obtained by LIBS and LIBS+LIF techniques are reported. The examined elements that were detected by LIBS in concentrations below the German regulatory values for unpolluted soils are: As, Cr, Cu, Ni, Pb, Ti and Zn. By applying LIBS+LIF, also Cd and Tl result detectable also in concentrations lower than those required for unpolluted soils. However, the LOD for Hg measured by LIBS is still well above the German regulatory limit (1 ppm) and further studies are necessary in order to lower its detection threshold.

Table 6.5. Limits of detection (LOD's) for heavy metals in soils [Hilbk-Kortenbruck et al 2001]

Element	Spectral line for LIBS detection (nm)	LOD by LIBS (ppm)	Spectral line for LIF detection (nm)	LOD by LIF (ppm)
As	235.0	3.3	-	
Cd	228.8	6	228.8	0.3
Cr	425.4	2.5	-	
Cu	324.8	3.3	-	
Hg	253.7	84	-	
Ni	231.6	6.8	-	
Pb	405.8	17	-	
Ti	351.9	48	-	
Tl	-	-	351.9	0.5
Zn	334.5	98	-	

6.5 Planetary Exploration

For the next decades, a series of robotic space exploration endeavors are planned to study several extraterrestrial bodies both by orbiters and landing missions. There is a particular interest in studying Mars, because it is the most Earth-like among the eight planets in the solar system. The main ambitions are to understand the present state of the planet, how it was formed, how it evolved and how its evolution compares with other planets including Earth. Furthermore, determining whether life ever existed or is even still active on Mars today is considered to be one of the major scientific questions in Mars science. In order to tackle these questions, it is of importance to determine the elemental composition of Martian surface rocks and soils. Presently there is only limited information on the chemical and mineralogical composition of the Martian surface, obtained from two Viking landers, Mars Pathfinder, from the presently active missions [Riedere et al. 1997, Golombek et al. 1997] and from the analysis of Martian meteorites [Edwards et al. 1999].

The future missions to Mars of NASA (Mars Science Laboratory) and ESA (Exomars) will include a robotic rover and instruments to analyze the elemental composition of the Martian surface materials by a LIBS technique. Due to very stringent requirements regarding the volume, weight and power consumption of the instruments for the planetary explorations, a LIBS instrument is of further interest as it is possible to achieve the system miniaturization and low power requirements [Del Bianco et al. 2006]. For given instrument characteristics, the experimental conditions must be optimized in order to obtain the maximum of analytical information on a variety of the samples (i.e. the detection of both major and trace constituents) and, contemporary, to minimize the power consumption. The optimization regards the number of laser shots to be applied and determination of the signal acquisition window in terms of delay from the laser pulse and gate duration [Colao et al. 2004, Sallè et al. 2005], which preferable captures the plasma while it is in LTE as it was discussed in sections 3 and 5.

A number of publications report quantitative LIBS analyses in simulated Mars atmosphere, obtained by different analytical approaches [Sallé et al. 2005 and 2006, Knight et al. 2000, Arp et al. 2004, Colao et al. 2004b]. Except one work [Arp et al. 2004], these analyses regard relatively dry samples and the measurements at room temperature.

6.5.1. Martian Environmental Conditions

Spacecraft- and Earth-based telescopic measurements mapped water vapor and ice-clouds in the Martian atmosphere [Jakosky et al. 2004]. The water vapor is transported by the winds, it can diffuse into the subsurface and so becomes water ice at high latitudes. The ice was mapped within near surface regions at high latitudes and on the surface in the polar regions. Spectroscopic measurements of sunlight reflected from the Mars surface indicate that about 1% (by mass) of water is adsorbed by soil, but also that water abundance varies by more than a factor of 10 from one year to the next. As a consequence, in LIBS analyses of Martian soils, the presence of water and ice should be taken into account. In this view, laboratory measurements on soil/ice mixtures have been already reported [Arp et al. 2004] at two different temperatures, namely 246 K and 165 K. It was found that the ablation rate (measured for one soil/ice mixture) and the signal was higher at higher sample temperature and for lower ice content.

The Martian atmosphere is composed mainly of CO_2 (95.3% by volume), N_2 (2.7%) and Ar (1.6%) with an average surface pressure of 6 mbar and seasonal pressure variations in the order of 30%. The Martian average surface temperature at the equator is about 220 K, however, strong temperatures variation exist between days an night, as well between different seasons. The European Space Agency (ESA) specifies that the LIBS instrument should operate at temperatures in the range -60°C to 30°C.

6.5.2. Behavior of Water and Ice in Soils at Low Temperatures

At a pressure of 6 mbar, water has triple point at 0.01 °C, and slightly below this pressure there is a direct transformation of water vapor into ice and vice-versa. Above the triple point, the water phase also exists.

At pressures below 2000 atm., and consequently also for low-pressure Mars environment, only Ice-I can be formed [Kou et al. 1993]. It can exist, depending on its temperature and formation, in three different modifications: hexagonal ice, cubic ice and vitreous ice. The latter can exists at temperatures bellow 150 K and warming leads to its irreversible transformation to the cubic ice. By warming above about 190 K, the cubic ice transforms into disordered, hexagonal ice.

However, when considering ice on the soil surface and inside moist soils, additional effects must be taken into account. External water on the surface of the grains nucleates to the normal form of hexagonal ice. Experimental measurements of ice films on a porous silica substrate, intended to simulate rock/ice interfaces as they appear in nature, show evidence of interfacial, supercooled water, down to temperatures of about -17°C [Engemann et al. 2004]. Inside the soils/rocks, so called pore ice exists, which has crystallite sizes of the same order as those of soil particles/pores [Dash et al. 1995]. Above the triple point, so also in air at atmospheric pressure, the unfrozen super-cooled water films can exist well below 0°C at interfaces with soil particles or at grain boundaries. Freezing of free pore water leads to the formation of a defective form of cubic ice (Ice-I) that is intermediate between hexagonal and cubic ice. The fraction of unfrozen water depends on sizes and shapes of the pores and grains, then on impurity concentrations and local environment (pressure and temperature). Depression of the water freezing point inside pores was measured by different authors and reviewed in [Schreiber et al. 2001]. The melting point of small ice crystals is inversely proportional to the crystal size i.e. pore size of the material, and super-cooled water inside pores was found to exist for the temperature down to 233 K. As for example, on Basalt, a small fraction of water was found down to about – 6°C, while at the same temperature a clay sample can contain also 20% of unfrozen water [Dash et al. 1995].

6.5.3. Temperature Dependence of the LIBS Signal from Moist Soils

Disordered surface ice, pore ice and super-cooled water have different physical properties [Kou et al. 1993], such as thermal conductivity, density, diffusion coefficient, viscosity, absorption coefficient, etc., and these properties can significantly affect the LIBS signal. While the surface ice, if present as a thin layer, can be removed by a certain number of laser shots before starting the actual LIBS measurements, at subzero temperatures water/ice inside pores is a part of the analyzed soil. In [Arp et al. 2004], the preparation of soil/ice mixture probably leads to the hexagonal ice formation and to quite different grain/ice interfaces than what we could expect in nature.

In following, the results of the first studies relative to the LIBS signal dependence of the soil temperature are presented [Lazic et al. submitted]. The measurements were performed on different types of soils and rocks in simulated Martian atmosphere (7 mbar pressure), in the temperature range from 25°C to -60°C. Both cooling and heating cycles were applied on the samples. The rock fragments had different grades of the surface polishing and the signal dependence on the surface roughness, which influences the melting temperature of the pore ice and interfacial ice, was investigated. During ablation at a single sample point, the spectrum was registered after each laser shot, allowing to eliminate signals from possible surface ice. For comparison, temperature dependent measurements were also performed on frozen water solution (disordered ice). On one soil sample, also the whole spectra were recorded at different temperatures and the reduction of the plasma electron density with lowering of the sample temperature was observed.

At subzero temperatures the LIBS signal shows more-less violent oscillations as a function of the sample temperature (Figures 6.11-6.13). This effect was attributed to the coo-existence of liquid-ice phase inside the rock/soil pores of different sizes, where the physical properties have a discontinuity. There are the indications that at these temperatures a part of the absorbed laser energy is lost to complete the phase transition before producing the temperature rise and successively the plasma. At certain temperatures and particularly for polished rock samples with more uniform pore size distribution, the measured LIBS signal depression was up to an order of magnitude (Figure 6.11). This leads to a great loss of the analytical information as many weak spectral lines can no more be distinguished from the noise. On irregular surfaces such as those usually found in nature, a larger distribution of the pore sizes inside the focal spot can be expected. Such distribution leads to a number of smaller negative peaks bellow 0°C referred to the signal dependence on the sample temperature, as it has been observed on the analyzed soil samples (not shown here) and on Figure 6.12.

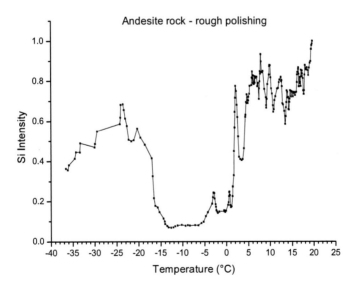

Figure 6.11. Normalized LIBS Si peak intensity (288 nm) as a function of the surface temperature: the rock sample was polished with a diamond wheel (grain size 70 μm); the spectra were accumulated over 40 laser shots and after applying 10 initial cleaning shots; cooling cycle, laser energy density 13.9 J/cm^2.

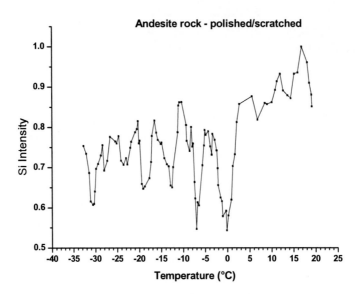

Figure 6.12. Normalized LIBS Si peak intensity (288 nm) as a function of the surface temperature; the rock sample was treated with an emery (grain size 140 μm); the signal was accumulated over 40 laser shots after applying 10 initial cleaning shots; cooling cycle, laser energy density 13.9 J/cm^2.

Disordered hexagonal ice is formed on the surface and this can be removed by laser ablation prior to the measurements. However, the same type of ice exist also inside larger pores and surface scratching, but during a relatively slow cooling cycles, as those expected in nature (also in Martian conditions), supercooled water can coexist with the ice phase. In the temperature range here examines (+30 °C to − 60°C), free supercooled water passes two transitions, one around -40°C and another about -50°C. These transitions were registered from depression of LIBS signal both on ice sample (not shown) prepared by slow cooling of water, and on moist rock/soil samples. The phase transition at lower temperature (-50 °C) produced the signal reduction also for 80% on one rock sample, while on the ice the signal was reduced even for two orders of magnitude. Very small ablation rate of the ice at this temperature can compromise also a possibility to remove the surface ice before analysis of underlying soil/rock surface.

Another depression point , more or less pronounced, was always registered on the soil/rock samples around 0°C although the eventual surface layer of the ice/water was previously removed by applying 10-30 laser shots. This effect was attributed to mixed presence of water and hexagonal ice inside larger pores or scratches.

Although here described measurements were performed in CO_2 environment at pressure of 7 mbar, similar effects can be expected in other environments above the triple point, such as in atmospheric air, whenever the moisture is present in the sample. From these results it can be suggested that LIBS analysis at subzero temperatures should exclude the operation close to the points of the water/ice transition, i.e. around 0°C, - 40°C and -50°C. The larger focal spot size would allow for interaction with a wider range of pore dimensions present in the natural soils/rocks, thus to reduce the signal changes with the temperature due to water/ice transitions dependent on the pore size.

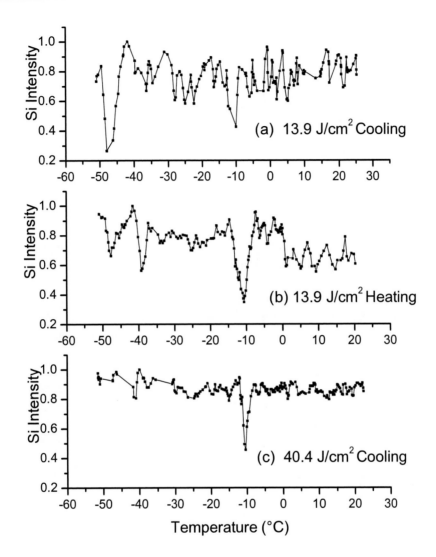

Figure 6.13. Normalized LIBS Si peak intensity (288 nm) as a function of the surface temperature; the rock was polished by a diamond (grain size 35 μm); the signal was accumulated by 40 laser shots and after applying 10 initial cleaning shots: (a) heating cycle and (b) cooling cycle at laser energy density of 13.9 J/cm^2; (c) cooling cycle, laser energy 40.4 J/cm^2.

7. CONCLUSION

The LIBS technique has been reviewed, with aim to supply to a reader the fundamentals regarding the instrumentations, physical processes and data analyses.

Different examples of LIBS applications are discussed, including the procedures for data elaboration and so obtained information about the analyzed samples, then the experimental aspects, the technique advantages and limitations for the particular cases.

Special emphasis is given on LIBS employment for soil/sediment analyses, being important for environmental monitoring and protection. Here, some LIBS applications different environments, are described in detail, including the experimental set-up used and the discussion of the obtained results. The influences of the sample physical and chemical properties, as well as of the local environmental conditions, are discussed and commented in a view of in-situ LIBS measurements.

For a reader interested in further information about LIBS technique and its applications, the relevant literature is cited.

ACKNOWLEDGMENTS

The author is deeply grateful to R. Fantoni, responsible of FIS-LAS unit (ENEA, Frascati) for the support in all her LIBS works.

A special thank is addressed to F. Colao (ENEA, Frascati) for the initial help with the LIBS instrumentation and LabView programming, and for early discussions relative to the data analyses.

A deep involvement of V. Spizzicchino (ENEA fellowship), and of S. Jovicevic (Institute of Physics, Belgrade), in different LIBS works here discussed is mentioned with gratitude.

The author is grateful to I. Rauschenbach (University of Munster) and to T. Santagata (CNR, Potenza), for the common measurements regarding LIBS for Mars exploration and LIBS for marble analyses, respectively.

Many thanks are also addressed to other colleagues from ENEA, Frascati, namely A. Palucci, L. De Dominicis, R. Giovagnoli, M. Carpanese and R. Ciardi, which on different ways, contributed to some of the reported works.

Except early work regarding Mars exploration, funded by ASI (Italian space agency), all the mentioned LIBS works from ENEA, Frascati, were funded by Italian Ministry for University and Scientific Research through different project (TECSIS, MIAO and PANDORA).

REFERENCES

Aguilera J.A. & Aragón, C. (2004). Characterization of a laser-induced plasma by spatially resolved spectroscopy of neutral atom and ion emissions. Comparison of local and spatially integrated measurements, *Spectrochim. Acta Part B*; 59, 1861-1876.

Aguilera, J.A., Aragón, C. & Begoechea, J. (2003). Spatial characterization of laser-induced plasmas by deconvolution of spatially resolved spectra, *Appl. Optics*; 42 (30), 5938-5946.

Amoruso, S., Bruzzese, R., Spinelli, N., et al. (1999). Characterization of laser ablation plasmas, *J. Phys. B;* 32, R131-R172.

Andor: http://www.andor.com/products/spectrographs/?product=ME5000.

Anzano, M., Villoria, M., Ruíz-Medina, A., et al. (2006). Laser-induced breakdown spectroscopy for quantitative spectrochemical analysis of geological materials: effects of the matrix and simultaneous determination, *Anal. Chim. Acta;* 575, 230-235.

Aragón, C., Peñalba, F. & Aguilera, J.A, (2005). Curves of growth of neutral atom and ion lines emitted by a laser induced plasma, *Spectrochim. Acta Part B*; 60, 879–887.

Aragón, C., Aguilera, J.A. & Penalba, F. (1999). Improvements in quantitative analyses of steel composition by Laser-Induced Breakdown Spectroscopy at atmospheric pressure using an infrared Nd:YAG laser, *Appl. Spectroscopy;* 53 (10), 1259-1267.

Arp, Z.A., Cremers, D.A., Wiens, R.C., et al. (2004). Analysis of water ice and water ice/soil mixtures using laser-induced breakdown spectroscopy: application to Mars exploration, *Appl. Spectrosc.* 58, 897-909.

Babushok, V.I., DeLucia, F.C. Jr., Gottfried, J.L, et al. (2006). Double pulse laser ablation and plasma: Laser induced breakdown spectroscopy signal enhancement, *Spectrochimica Acta Part B;* 61, 999–1014.

Baghdassarian O., Tabbert, B., Williams, G. A., et. al. (1999) Luminescence Characteristics of Laser-Induced Bubbles in Water, *Phys. Rev. Lett.*, 83, 2437- 2440.

Balzer, H., Hoehne, M., Sturm, V., et al. (2005). Online coating thickness measurement and depth profiling of zinc coated sheet steel by laser-induced breakdown spectroscopy, *Spectrochimica Acta Part B 60,* 1172 – 1178.

Barbini, R., Colao, F., Lazic, V., et al. (2002). On board LIBS analysis of marine sediments Collected during the XVI Italian campaign in Antartica, *Spetrochim. Acta Part B*; 57, 1203-1218.

Body, D. & Chadwick, B.L. (2001). Optimization of the spectral data processing in a LIBS simultaneous elemental analyses system, *Spectrochim. Acta B*; 56, 725-736.

Boumans, P.W.J.M. (1976). Excitation of spectra, in Comprehensive Analytical Chemistry (G. Svehla Ed.), Wilson-Willson's, Elsevier Pub. Co, Amsterdam, p. 1-193.

Brambati, A., Fanzutti, G.P, Finocchiaro, F., et al. (2000). Some palaeoecological remarks on the Ross Sea Shelf, in (F. M. Faranda et. al. Ed), Ross Sea Ecology. Springer, Berlin Heidelberg, 51-61.

Bublitz, J., Dölle, C., Schade, W., et al. (2001). Laser-induced breakdown spectroscopy for soil diagnostics, *Eu. J. of Soil Sci.* 52, 305-312.

Bulatov, V., Xu, L. & Schechter, I. (1996). Spectroscopic Imaging of Laser-Induced Plasma, *Anal. Chem.* 68, 2966-2973.

Carranza, J.E. & Hahn, D.W. (2002). Assesment of the upper particle size limit for quantitative analyses of aerosols using laser induced breakdown spectroscopy, *Anal. Chem.* 74, 5450-5454.

Carranza, J.E, Gibb, E., Smith, B.W, et al. (2003). Comparison of nonintensified and intensified CCD detectors for laser-induced breakdown spectroscopy, *Appl. Optics;* 42, 6016-6021.

Castle, B.C., Visser, K., Smith, B.W., et al. (1997). Spatial and temporal dependence of lead emission in laser-induced breakdown spectroscopy, *Appl. Spectrosc.* 51 (7), 1017-1024.

Ceccaroni, L., Frank, M., Frignani, M., et al. (1998). Late Quaternary fluctuations of biogenic component fluxes on the continental slope of the Ross Sea, Antarctica, *J Mar Syst.* 17, 515-525.

Chaleard, C., Mauchien, P., Andre, N., et al. (1997). Correction of Matrix effects in Quantitative Elemental Analyses with Laser Ablation Optical Emission Spectrometry, *J. Anal. At. Spectrom.* 12, 183-188.

Colao, F., Fantoni, R., Lazic, V. et al. (2002). Laser Induced Breakdown Spectroscopy for semi-quantitative analyses of artworks - application on multi-layered ceramics and copper based alloys, *Spetrochim. Acta B;* 57, 1219-1234.

Colao, F., Lazic, V. & Fantoni, R. (2002b). LIBS as an analytical technique in non equilibrium plasmas, SPIE Special Issue on "Spectroscopy of non equilibrium plasma at elevated pressures." Vol.4460 (Ockin V. N. Ed) 339-348.

Colao, F., Lazic, V., Fantoni, R., et al. (2004). LIBS application for analyses of martian crust analogues: search for the optimal experimental parameters in air and CO2 atmosphere, *Appl. Phys. A;* 79, 143-152.

Colao, F., Fantoni, R., Lazic, V., et al. (2004b). Investigation of LIBS feasibility for in situ planetary exploration: An analysis on Martian rock analogues, *Planetary and Space Sci.* 52, 117-123.

Collier, R., Dymond, J., Honjo, S., et al. (2000). The vertical flux of biogenic and lithogenic material in the Ross Sea: moored sediment trap observations 1996-1998, *Deep-Sea research II*, 47, 3491-3520.

Corsi, M., Cristoforetti, G., Palleschi, V., et al. (2001). A fast and accurate method for the determination of precious alloy carratage by Laser induced Plasma Spectroscopy, *Eur. Phys. J.,* D13, 373-377.

Cremers, D.A. & Radziemsky, L.J. (2006). Handbook of Laser Induced Breakdown Spectroscopy, John Wiley & Sons Ltd.

Dash, J.G., Fu, H. & Wettlaufer, J.S. (1995). The premelting of ice and its environmental consequence, *Rep. Prog. Phys.* 58, 115-167.

Dehairs, F., Stroobants, N. & Goeyens, L. (1991). Suspended barite as a tracer of biological activity in the Southern Ocean, *Marine Chemistry*, 35, 399-410.

Del Bianco, A., Rauschenbach, I., Lazic, V., et al. (2006). GENTNER – a miniaturised LIBS/Raman instrument for the comprehensive in situ analysis of the Martian surface, 4th NASA International Planetary Probe Workshop.

De Lucia, F.C., Harmon, R.S., McNesby, K.L. et al. (2003). Laser-induced breakdown spectroscopy analysis of energetic materials, *Appl. Optics,* 42, 6148- 6152.

De Mater, D.J. (1981). The supply and accumulation of silica in the marine environment, *Geochim. Cosmochim. Acta*; 45, 1715-1732.

Edwards, H.G.M., Farwell, D.W., Grady, M.M., et al. (1999). Comparative Raman microscopy of a Martian meteorite and Antarctic lithic analogues, *Planetary Space Sci.* 47, 353-362.

Engemann, S., Reichert, H., Dosch, H., et al. (2004). Interfacial melting of ice in contact with SiO2, Phys. Rev. Lett., 92 (20) 205701: 1-4.

Eppler, A.S., Cremers, D.A., Hickmott, D.D., et al. (1996). Matrix effects in the detection of Pb and Ba in soils using laser-induced breakdown spectroscopy, *Appl. Spectrosc.* 50, 1175 – 1181.

Florek, S., Haischa, C., Okruss, M., Becker-Ross, H., (2001) A new, versatile echelle spectrometer relevant to laser induced plasma applications, *Spectrochim. Acta Part B*, 56, 1027-1034.

Fornarini, L., Colao, F., Fantoni, R., et al. (2005). Calibration for LIBS analyses of bronze materials by nanosecond laser excitation: a model and an experimental approach, *Spectrochim. Acta B;* 60, 1186-1201.

Giakoumaki, A., Melessanaki, K. & Anglos, D. (2007). Laser-induced breakdown spectroscopy (LIBS) in archaeological science - applications and prospects, *Anal. Bioanal. Chem.* 387, 749-760.

Golombek, M.P., Cook, R.A., Economou, T., et al. (1997). Overview of the Mars Patyhfinder Mission and assessment of landing site prediction, *Science* 278, 1743-1748.

Gornushkin, I. B., Smith, B.W., Omenetto, N., et al. (2004). Microchip laser-induced breakdown spectroscopy: preliminary feasibility study, *Appl. Spectrosc.* 58, 763-769.

Griem, H. R. (1964). *Plasma Spectroscopy*, McGraw-Hill, USA.

Griem H.R. (1974). Spectral line broadening by plasmas, Academic press NY and London.

Hamilton, S. Al-Wazzan, R., Hanvey, A., et al. (2004). Fully integrated wide wavelength range LIBS system with high UV efficiency and resolution, *J. of Anal. At. Spectrom.* 19, 479–482.

Hilbk-Kortenbruck, F., Noll, R., Wintjens, P., (2001). Analysis of heavy metals in soils by using laser-induced breakdown spectrometry combined with laser-induced fluorescence, *Spectrochim. Acta Part B*, 56, 933-945.

Huddlestone, R. G., Leonard, S. L., (1965), Plasma diagnortics techniques ,Academic Press, New York, p. 209.

Jakosky, B.M & Mellon, M.T (2004). Water on Mars, *Physics Today*, April 71-85.

Knight, A.K, Scherbarth, N.L, Cremers, D.A, et al. (2000). Characterization of Laser-Induced Breakdown Spectroscopy for application to space exploration, *Appl. Spectrosc.* 54, 331-340.

Konjevic, N. (1999). Plasma broadening and shifting of non-hydrogenic spectral lines: present status and applications, *Phys. Reports* 316, 339-401.

Kou, L, Labrie, D. & Chylek, P. (1993). Refractive indices of water and ice in the 0.65- to 2.5-μm spectral range, *Appl. Optics* 32 (19), 3531-3540.

Kumar, N., Anderson, R.F., Mortlock, R.A, et al. (1995). Increased biological productivity and export production in the glacial. *Southern Ocean, Nature;* 378, 675-680.

Kurucz data base, available at: http://cfa-www.harvard.edu/amdata/ampdata/kurucz23/sekur.html.

Lazic, V., Barbini, R., Colao, F., et al. (2001). Self absorption model in quantitative Laser Induced Breakdown Spectroscopy measurements on soils and sediments, *Spectrochim. Acta B;* 56, 808-820.

Lazic, V., Fantoni, R., Colao, F., et al. (2004). Quantitative Laser Induced Breakdown Spectroscopy analysis of ancient marbles and corrections for the variability of plasma parameters and of ablation rate, *J. Anal. At. Spectrom.* 19, 429-436.

Lazic, V., Colao, F., Fantoni, R., et al. (2005). Laser Induced Breakdown Spectroscopy in water: improvement of the detection threshold by signal processing, *Spectrochim. Acta Part B;* 60, 1002-1013.

Lazic, V., Colao, F., Fantoni, R., et al. (2005b). Recognition of archeological materials underwater by laser induced breakdown spectroscopy, *Spectrochim. Acta B,* Vol 60, 1014-1024.

Lazic, V., Colao, F., Fantoni, R., et al. (2007). Underwater sediment analyses by laser induced breakdown spectroscopy and calibration procedure for fluctuating plasma parameters, *Spectrochim. Acta Part B;* 62, 30–39.

Lazic, V., Rauschenbach, I., Jovicevic, S., et al. (2007). Laser induced breakdown spectroscopy of soils, rocks and ice at subzero temperatures in simulated martian conditions, *Spectrochim. Acta Part B*, 62, 1546–1556.

Lochte-Holtgraven, W. (1995). Plasma diagnostics, AIP, New York.

Lohse, D., (2002). Sonoluminescence: inside a microreactor, *Nature*, 418, 381-383.

López-Moreno, C., Palanco, S. & Laserna, J.J. (2004). Remote laser-induced plasma spectrometry for elemental analysis of samples of environmental interest, *J. Anal. At. Spectrom.* 19, 1479 – 1484.

Martin, M.Z., Wullschleger, S.D., Garten, C.T. Jr, et al. (2003). Laser-induced breakdown spectroscopy for the environmental determination of total carbon and nitrogen in soils, *Appl. Optics;* 42, 2072-2077.

Mao, S.S., Zeng, X., Mao, X., et al. (2004). Laser-induced breakdown spectroscopy: flat surface vs. cavity Structures, *J. Anal. Atom. Spectrometry*; 19, 1295–1301.

Multari, R.A, Forster, L.E., Cremers, D.A., et al. (1996). The effects of sampling geometry on elemental emissions in laser-induced breakdown spectroscopy, *Appl. Spectrosc.* 50, 1483-1499.

NIST electronic database, available at http://physlab.nist.gov/PhysRefData/contents-atomic.html.

Piepmeier, E.H. (1986). Laser Ablation for Atomic Spectroscopy, in E.H. Piepmeier (Ed.), Analytical Application of Laser. John Wiley & Sons (New York).

RadziemskiL, J. (2002). From LASER to LIBS, the path of technology development, *Spectrochim. Acta Part B*; 57, 1109-1113.

Rieder R, Economou T, Wänke H, Turkevich A, Crisp J, Brückner J, Dreibus G, McSween H. Y Jr., 1997, The chemical composition of Martian soil and rocks returned by the mobile alpha proton X-ray spectrometer: preliminary results from the X-ray mode, Science 378, 1771-1774.

Saggiomo, V., Carrada, G.C, Mangoni, O., et al. (2000). Ecological and physiological aspects of primary production in the Ross Sea, in (F. M. Faranda et. al. Ed), Ross Sea Ecology. Springer, Berlin Heidelberg, 247-258.

Sallé, B., Cremers, D.A, Maurice, S., et al. (2005). Laser-induced breakdown spectroscopy for space exploration applications: influence of the ambient pressure on the calibration curves prepared from soil and clay samples, *Spectrochim. Acta Part B* 60, 479-490.

Sallé, B., Cremers, D.A, Maurice, S., et al. (2005b). Evaluation of a compact spectrograph for in-situ and stand-off laser-induced breakdown spectroscopy analyses of geological samples on Mars missions, *Spectrochim. Acta Part B;* 60, 805-815.

Sallé, B., Lacour, J.L., Mauchien, P., et al. (2006). Comparative studies of different methodologies for quantitative rock analysis by Laser-Induced Breakdown Spectroscopy in a simulated Martian atmosphere, *Spectrochim. Acta Part B*; 61, 301-313.

Schreiber, A., Ketelsen, I. & Findenegg, G.H. (2001). Melting and freezing of water in ordered mesoporius silica materials, Phys. Chem. *Chem. Phys.* 3, 1185-1195.

Sdorra, W. & Niemax, K. (1992). Basic investigations for laser microanalysis: III. Application of different buffer gases for laser-produced sample plumes, Mikrochim. Acta 107, 319-327.

Siegel, J., Epurescu, B., Perea, A., et al. (2005). High spatial resolution in laser-induced breakdown spectroscopy of expanding plasmas, *Spectrochim. Acta Part B*; 60, 915-919.

St-Onge, L., Sabsabi, M. & Cielo, P. (1997). Quantitative Analyses of Additives in Solid Zinc Alloys by Laser-Induced Plasma Spectrometry, *J. Anal. At. Spectrom.* 12, 997-1004.

Stellarnet Inc: http://www.stellarnet-inc.com/.

Tognoni, E, .Palleschi, V., Corsi, M., et al. (2002). Quantitative micro-analyses by laser-induced breakdown spectroscopy: a review of the experimental approaches, *Spectrochim Acta B;* 57, 1115-1130.

Van der Mullen, J.A.M. (1990). Excitation equilibria in plasmas; a classification, *Physics Reports* 191, 2&3 109-220.

Zhang, X., Chu, S.S..,Ho, J.R, et al. (1997). Excimer laser ablation of thin gold films on a quartz crystal microbalance at various argon background pressures, *Appl. Phys. A;* 64, 545-552.

Yalcin, S., Crosley, D.R., Smith, G.P., et al. (1999). Influence of ambient conditions on the laser air spark, Appl. Phys. B 68, 121-130.

Wainner, R.T. Harmon, R.S., Miziolek, A.W, et al. (2001). Analysis of environmental lead contamination: comparison of LIBS field and laboratory instruments, Spectrochim. Acta Part B 56, 777-793.

Winefordner, D, Gornushkin, I.B, Correll, T, et al. (2004). Comparing several atomic spectrometric methods to the super stars: special emphasis on laser induced breakdown spectrometry, LIBS, a future super star, J. Anal. At. Spectrom., 19, 1061-1083.

Wisbrun, R. Schechter, I., Niessner R. et al. (1994). Detector for trace elemental analysis of solid environmental samples by laser plasma spectroscopy, *Anal. Chem.* 66, 2964-2975.

Zayhowski, J. J. (2000). Passively Q-switched Nd:YAG microchip lasers and applications, *J. of Alloys and Compounds*, 303–304, 393–400.

In: Laser Applications in Environmental Monitoring
Editors: L. Fiorani and F. Colao

ISBN 978-1-60456-249-1
© 2008 Nova Science Publishers, Inc.

Chapter 3.2

THREE-DIMENSIONAL SCAN OF UNDERGROUND CAVITIES

Roberto Ricci

Laser Applications Section, Advanced Physical Technologies and New Materials
Department, Italian National Agency for New Technologies, Energy
and the Environment (ENEA),
via Enrico Fermi 45, 00044 Frascati, Italy
E-mail: roberto.ricci@frascati.enea.it, tel.: +39 – 06-94 00 55 43

ABSTRACT

Laser rangefinding has recently become an important method for distance measurements. In particular, laser rangefinders (LRFs) are being increasingly used for rapid 3D digitization of single objecs or entire scenes. The process consists of scanning a physical target by a laser scanner which gives information about its geometry and results in an accurate 3D computer model or "fingerprint" of the target. Possibly combined with digital photos that provide colour and texture information, the 3D model can be used in virtual environments, to create replicas i.e. rapid manufacturing and prototyping and for scientific purposes in various application fields, including cultural heritage preservation and investigation of underground archaeological sites. We report here a brief survey of the most important methods for 3D digitization, with an emphasis on laser rangefinding techniques based on amplitude modulation. Some examples of applications of 3D digitization techniques in underground environments are also given.

INTRODUCTION

Distance measuring devices and corresponding methods to digitize and reconstruct the shapes of complex three dimensional real targets have evolved rapidly in recent years. The speed and accuracy of digitizing technologies owe much to advances in the areas of physics and electrical engineering, including the development of lasers, CCD's, and high speed sampling and timing circuitry. Considerable progress has been achieved in understanding the basic theoretical and practical principles of range sensing and it is to be expected that the advance in the design of lasers, integrated optics devices, emitter and receiver electronics as well, will lead to further interesting developments.

Applications range from exploration of extreme or even hostile environments to cultural heritage cataloguing and conservation, non-destructive testing, reverse engineering, 3D vision and virtual reality. In particular, non-contact distance measurements of inaccessible or fragile targets are of great interest for many scientific and industrial applications. The basic principle of active non-contact rangefinding devices is to project a signal (radio, ultrasonic or optical) onto an object and to process the reflected or scattered signal to determine the distance. If a high resolution rangefinder is needed, an optical source – usually a laser - must be chosen because radio and ultrasonic waves cannot be focused adequately. Based on the range sensing technique used, laser rangefinders (LFR's) are further traditionally organised into four categories: interferometric, pulsed time-of-flight, triangulation and amplitude modulation devices.

In this work, we address methods of both digitizing and reconstructing the shapes of complex targets, by firstly presenting a tentative taxonomy of the variety of range acquisition methods available. Then we focus on laser rangefinders, briefly analysing advantages and limitations of each technique as compared to the others. We finally describe in greater detail the functioning principle and characteristics of amplitude modulation LRF's, with an emphasis on applications in underground environments.

OVERVIEW OF RANGEFINDING TECHNIQUES

A vast number of range acquisition methods have evolved over the last decades. These methods follow two primary directions: passive and active sensing.

Passive approaches do not interact with the object, whereas active methods make contact with the object or project some kind of energetic probe onto it. The computer vision research community has been largely focused till very recently on passive methods that extract shape from one or more digitized 2D images. These methods do not need their own light source, but they use the ambient light for gathering the distance information from the target. Computer vision approaches include shape-from-shading for single images, stereo triangulation for pairs of images, and optical flow and factorization methods for video streams. While these methods require very little special purpose hardware, they typically do not yield the dense and highly accurate digitizations that are required of a number of applications.

More interesting for the purposes of this work are active range acquisition methods. Figure 1 reports a tentative taxonomy, which is by no means comprehensive, but rather intended to introduce the reader to the variety of methods available. Among active sensing methods, two main different approaches can be distinguished: contact and non-contact sensors.

Contact sensors are typically touch probes that consist of jointed arms or pulley-mounted tethers attached to a narrow pointer. The angles of the arms or the lengths of the tethers indicate the location of the pointer at all times. By touching the pointer to the surface of the object, a contact event is signalled and the position of the pointer is recorded. Touch probes come in a wide range of accuracies as well as costs. Coordinate Measuring Machines

(CMM's) are extremely precise (and costly), and they are still widely adopted in industrial manufacturing for making precision shape measurements.

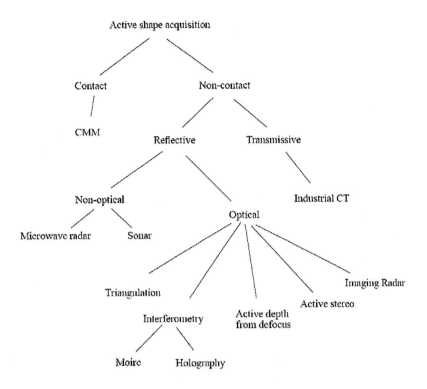

Figure 1. Tentative taxonomy of active range acquisition methods.

Though very accurate, touch probes have several drawbacks:

- they are slow;
- they can be clumsy to manipulate;
- they usually require to be supervised by a human operator;
- they must make contact with the surface, which may be undesirable for fragile objects;
- they cannot be used for the digitization of entire scenes, especially in hardly accessible environments such as underground cavities.

Active, non-contact methods generally operate by projecting energy waves onto the target, followed by recording the transmitted or - more usually - reflected energy. A powerful transmissive approach for shape capture is for example industrial computer tomography (CT). Industrial CT entails bombarding an object with high energy x-rays and measuring the amount of radiation that passes through the object along various lines of sight. After back projection or Fourier projection slice reconstruction, the result is a high resolution volumetric description of the density of space in and around the object. This volume is suitable for direct visualization or surface reconstruction. The principal advantages of this method over reflective methods are its insensitivity to the reflective properties of the surface, and its ability to capture the internal cavities of an object that are not visible from the outside. The principal disadvantages of industrial CT scanners are:

- they are very expensive;
- large density variations due to change of material composition can degrade accuracy;
- they are potentially very hazardous due to the use of radioactive materials;
- they are well suited to the acquisition of volumetric information about self-contained objects, but of little use for distance survey applications.

Active, reflection methods for shape acquisition can be further subdivided into two categories: non-optical and optical approaches. Non-optical approaches include sonar, radio frequency and microwave radar (Radio Detecting and Ranging), which typically measure distances to an object by measuring the time required for a pulse of sound or electromagnetic energy to bounce back from the object itself. Amplitude or frequency modulated continuous energy waves can also be used in conjunction with phase or frequency shift detectors.

Sonar range sensors are typically inexpensive, but they are also not very accurate and do not have high acquisition speeds. Microwave radar is typically intended for use with long range remote sensing, though close range optical radar is feasible, as described below.

The properties of range finders using gigahertz range radio frequency waves and optical methods have been compared in [Hovanessian 1988]. A common factor to both range finder types is that the angle resolution depends on wavelength. The divergence of the beam decreases, when the wavelength decreases. The optical methods are the best when a high resolution in horizontal and vertical dimensions is required. As the optical methods also enable the best resolution and accuracy in distance measurement, they are the most suitable methods in producing an accurate 3D image from the target, if the medium used (usually air) has not too much particles, like dust or smoke in it.

The last category in our taxonomy consists of active, optical reflection methods. The desire to capture shape by optical means dates back to the beginning of photography. In the 1860s a process was invented known as photosculpture, which used 24 cameras. Profiles of the subject to be reproduced were taken on photographic plates, projected onto a screen using the magic lantern and transferred to a piece of clay using a pantograph. Commercial applications developed rapidly and stayed in operation for a few years, when it was realized that the photosculpture process was not more economical than the traditional way of doing sculpture. In any case, the process still required a lot of human intervention. It is only with the advent of computers and lasers that the process has regained substantial interest, more than 100 years later.

In optical reflection methods, light – possibly structured - is projected onto an object and shape determined by measuring the light (back-) reflected from the object. In contrast to passive and non-optical methods, active optical rangefinders are suited to rapidly acquire dense, highly accurate range samplings. In addition, they are safer and less expensive than industrial CT, with the obvious limitation that they can only acquire the optically visible portions of the surface. These optical methods include active depth-from-defocus, active stereo and laser rangefinding methods - such as active triangulation, interferometry, pulsed time-of-flight and phase-shift or amplitude modulated rangefinding – which are presented in deeper detail in the following section.

Active depth from defocus operates on the principle that the image of an object is blurred by an amount proportional to the distance between points on the object and the in-focus object plane. The amount of blur varies across the image plane in relation to the distances of the imaged points. This method has evolved as both a passive and an active sensing strategy. In the passive case, surface reflectance variations (also called surface texture) are used to determine the amount of blurring. Thus, the object must have surface texture covering the whole surface in order to extract shape. Further, the quality of the shape extraction depends on the sharpness of surface texture. Active methods avoid these limitations by projecting a pattern of light (e.g., a checkerboard grid) onto the object. Most prior work in active depth from defocus has yielded moderate accuracy (up to one part per 400 over the field of view [Nayar 1995]).

Active stereo uses two or more cameras to observe features on an object. If the same feature is observed by two cameras, then the two lines of sight passing through the feature point on each camera's image plane will intersect at a point on the object. As in the depth from defocus method, this approach has been explored as both a passive and an active sensing strategy. Again, the active method operates by projecting light onto the object to avoid difficulties in discerning surface texture.

LASER RANGEFINDING TECHNIQUES

Laser rangefinding has become an important method for distance measurements recently and laser rangefinders are being increasingly used for rapid 3D digitization. The process consists of scanning a physical object by a laser scanner which gives information about the geometry of the object and results in a 3D fingerprint' of the object. Possibly combined with digital photos of the object that provide colour and texture information, the 3D model can be used in virtual environments, special effects for the film industry, video gaming, and can also be used for remote inspection or to create replicas i.e. rapid manufacturing and prototyping. Today commercial scanners are available that provide several thousands scan points per second with less than 1mm accuracy and that can operate to distances up to several tens of meters by meeting laser safety requirements at the same time. Some examples of 3D digitization are given below:

- LRF's can be used to rapidly acquire the geometry of building façades in a city. Combined with video images they can be used to create 3D photorealistic city models.
- LRF's are used to digitize precious works of art such as the works of Michelangelo, http://graphics.stanford.edu/projects/mich/, and the Statue of Liberty, http://www.geomagic.com/advantage/graphics/ liberty-index.php3.
- LRF's mounted on an airplane are used for example by geographical survey services to build digital elevation models and digital terrain maps. In this context the process is also known as LIDAR, an acronym standing for Light Detection and Ranging [see the contributions of V. Mitev and T. Kearns and L. MacDonald in this book].
- LRF's are used by entertainment studios to digitize humans for use in films, video games, special effects etc.

Besides 3D digitization of objects, LRFs can be used as sensors for robotic navigation, obstacle and collision avoidance, inspection, exploration of hardly accessible environments etc.

The basic principle of LRF's, as for other active optical methods, is to project an energy probe – specifically, a laser beam - onto an object and to process the reflected or scattered light to determine the distance. The use of lasers in this context over other forms of signals is motivated by the following reasons:

- it is easy to build compact systems that can produce highly focussed, low divergence beams. Radio and ultrasonic waves cannot be focussed adequately.
- A laser beam is bright and, because of low divergence, stays bright for large distances.
- Monochromaticity of the laser beam generally leads to simpler signal processing architectures.

In the following subsections we consider various methods for laser rangefinding such as interferometry, pulsed time-of-flight, triangulation and amplitude modulation. In all these methods the light projection is most commonly done via a laser diode and associated optics such as beam splitters and lenses etc. The laser diode converts electrical current into light wave or pulse. The reflected signal is detected with an avalanche photodiode (APD) and associated optics such as lenses for focussing the reflected signal. The APD converts light wave or pulse into electrical current. Amplifiers are used to suitably amplify the signal and appropriate signal processing electronics such as filters are used to reject noise, shape the signal, perform calculations etc.

Interferometry

Interferometric methods operate by projecting a spatially or temporally varying periodic pattern onto a surface, followed by mixing the reflected light with a reference pattern. The reference pattern demodulates the signal to reveal the variations in surface geometry. Interferometric methods include Moiré techniques, holographic interferometry and Fresnel diffraction.

Moiré interferometry involves the projection of coarse, spatially varying light patterns onto the target, which is typically illuminated with laser light through a grating (see Figure 2). The image of the target is captured through another grating with a camera. The gratings create a diffraction pattern and the distance can be calculated by measuring the position of diffraction stripes.

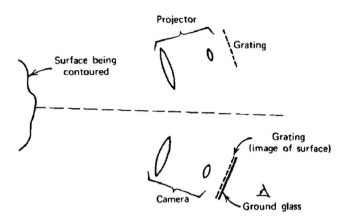

Figure 2. Typical experimental setup for Moiré interferometry.

Holographic interferometry typically relies on mixing coherent illumination with different wave vectors. A hologram is created from the target and the hologram is used as a reference object. When the hologram and the original target are illuminated with a laser beam, their combination creates an interference pattern. The distance can be calculated from the distance between the lines in the pattern.

The Fresnel diffraction is based on illuminating a grating with coherent light. The grating creates interference patterns at regular distances, which can be used for distance measurement.

In general, interferometers are very accurate, even in μm-class. Moiré methods can have phase discrimination problems when the surface does not exhibit smooth shape variations. This difficulty usually places a limit on the maximum slope the surface can have to avoid ranging errors. Holographic methods typically yield range accuracy of a fraction of the light wavelength over microscopic fields of view. A limitation is that without reference distances it is impossible to measure distances unambiguously, and only the differences in distance can be measured. Using several wavelengths also absolute measurements can be carried out [Bien 1981]. One application of interferometry is to measure the movements of earth's crust [Vali 1966].

Pulsed Time-of-Flight

Time-of-flight (ToF) radars are essentially the same as microwave radars, apart from operating at optical frequencies. The functioning principle is very simple (see e.g. [Goldstein 1967]. A laser pulse is projected onto the scene and the time the pulse takes to hit the target, reflect back and reach the detector is measured. If d is the distance to the target, t is the echo time, and c is the speed of light, then, trivially:

$$2d = ct$$

Since for a non-ambiguous measurement t should be greater than the pulse width T_p ($t > T_p$), it follows that

$$d > \frac{1}{2}cT_p$$

For a typical laser pulse $T_p = 10$ ps, this yields d > 1.5 mm. Practically, all the effort in designing a ToF LRF lies in measuring the echo time t accurately. The error in distance estimation is directly proportional to t,

$$\delta d = \frac{1}{2}c\delta t$$

So an error of 1ps gives, for example, $\delta d = 0.15$mm.

ToF LRFs are among the most commonly used systems for 3D digitization. They are usually available as 2D or 3D scanners. The general scheme of a ToF range finader is shown in Figure 3. Sweeping the laser beam by a rotating mirror a 2D scanner is obtained, which is able to provide scan points that lie in the plane in which the laser beam lies. A good representative of current technology is the LMS scanner available from SICK, http://www.sick.com, which does a 180 degree sweep of the beam with 1 degree angular resolution every 13ms resulting in a scan rate of approximately 13846 points per second. If the scanning plane is rotated at regular time intervals about an axis, a 3D scanner is obtained. A good representative is the DeltaSphere-3000 3D Digitizer from 3rdTech, http://www.3rdtech.com.

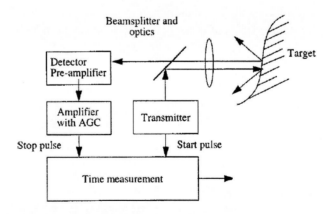

Figure 3. General scheme of a ToF range finder.

The time discriminator is a very important part of the precision time measurement system in the ToF LRF. The task of the discriminator is to observe time information from the electric pulse of the detector preamplifier and to produce a triggering signal at the right instant. Methods and algorithms used to estimate the pulse's return time depend on the desired time resolution, counting rate and required dynamic range. Commonly used principles in discriminator design include leading edge timing (constant amplitude), zero crossing timing (derivation), first moment timing (integration), and constant fraction timing. Constant Fraction Discriminator (CFD) is the more popular. The principle behind the operation of CFD is the search for an instant in the pulse when its height bears a constant ratio to pulse amplitude. The occurrence of this point produces a triggering pulse. The time interval between the start and stop pulses is measured with the time-to-digital converter (TDC), which is a fast, accurate and stable time-interval measuring device that uses e.g., a digital counting technique together with an analogue or digital interpolation method.

The main sources of inaccuracy in ToF laser rangefinders are noise-generated timing jitter, walk, nonlinearity and drift. Typical noise sources include noise generated by the electronics, shot noise caused by the background radiation induced current and shot noise created by the noise of the signal current. Jitter in timing determines mainly the precision of the range measurement. A thorough discussion of these aspects can be found in [Amann 2001].

For large objects, a variety of ToF radars have been demonstrated to give excellent results. For smaller objects, on the order of one meter in size, attaining 1 part per 1000 accuracy with a ToF radar requires very high speed timing circuitry, because the time differences to be detected are in the femtosecond (10^{-15} s) range.

Several applications of pulsed ToF laser range finders have been presented in [Riegl 2001]. The applications have been classified to surveying, measurement of height, profiling, industrial applications and traffic safety. In these applications the distance varies in a large range, from zero to kilometres and the accuracy required varies from some centimetres to some tens of centimetres with passive targets. For example in surveying and in localizing the exact position of targets in buildings and in accurate robot work even millimetre level accuracy may be required. On the other hand, in warning systems and in long distances the accuracy required can be some centimetres, and in military applications even some metres. In real time control of robots and industrial processes and modelling of 3D targets also high measuring speed is needed, however in most applications speed is not critical, a measuring time of some seconds for one point is enough.

The first applications in the 1960s were measurements of height or surveying, often over long distances, especially in space to satellites or to the moon [Bender 1967]. One of the first tests was also a laser range finder using a semiconductor laser and a light multiplying tube, which was used for measuring distances to passive targets in distances of some hundred metres to passive targets and even kilometres to active targets [Goldstein 1967].

At shorter distances the applications of surveying are, for example, measuring the shape of walls and roofs of houses, which is necessary for defining the exact positions and dimensions of old buildings. Measuring of height is useful in defining the position of aeroplanes, missiles and parachutes.

One of the most popular laser range finder applications has been defining the profile of the target and identifying it, e.g. a car passing under the range finder. Also measuring the shape and volume of underground caves and tunnels or overground piles of materials is a common profiling applications. The differences in height of earth ground have been measured from aeroplanes using semiconductor or Nd:YAG-lasers [Bufton 1989].

Some examples of industrial applications are positioning of ships and cranes, measuring the amount of stored liquids and solid materials, proximity switches, positioning of machine tools and measuring the tension of conveyor belts. One industrial application is also to measure the changes in the shapes of constructions by integrating an optical fibre inside the construction, which can be a glass-fibre container or a steel bar in a bridge, for example. It is possible to measure even local changes in the constructions, because the fibre contains Bragg grating reflectors positioned at even intervals [Lyöri 1997]. ToF laser range finders can also be used in traffic control and in safety control of automatically moving vehicles.

In 3D vision the ToF method gives several advantages compared to other optical methods. The measurement result is unambiguous and accurate both in horizontal and vertical directions, because the measurement beam is narrow and there is no danger of the target being hidden behind some other object in the scene, like in triangulation. A further advantage of the laser beam is that it can be used, for example, to help direct a CCD or CMOS camera. The ToF method is also often faster than the methods based on analyzing the image created with a camera.

Optical Triangulation

Optical triangulation is one of the most popular optical rangefinding approaches, also due to the availability of several commercial LRF's based on optical triangulation at relatively low cost. Figure 4 shows a typical system configuration in two dimensions.

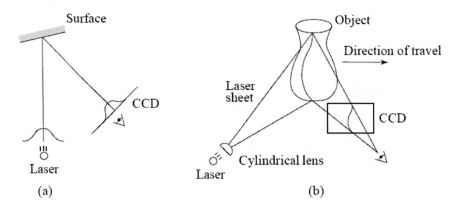

Figure 4. Optical triangulation and range imaging. (a) In 2D, a narrow laser beam illuminates an object's surface, and a CCD sensor images the reflection from the object. The centre of the image pulse maps to the centre of the laser, yielding a range value. (b) In 3D, a laser stripe triangulation scanner first spreads the laser beam into a sheet of light with a cylindrical lens. The CCD observes the reflected stripe from which a depth profile is computed. The object sweeps through the field of view, yielding a range image. Other scanner configurations rotate the object to obtain a cylindrical scan or sweep a laser beam or stripe over a stationary object.

In this method we have a laser projector projecting a laser beam (or stripe, see below) into the scene and a camera viewing the scene at a certain angle. Both the camera and the projector have a local coordinate system associated with them. The distance between the camera and the projector is termed the baseline. After hitting a point P on an object, the laser beam is scattered and an image of P is recorded by the camera. If 1) the camera type of projection (perspective projection being most common), 2) the intrinsic camera calibration matrix (see for example [Ma 2003]), that takes care of parameters such as the camera focal length and pixel size, and 3) the coordinates of the centre of projection of the camera are known, then given the image point p we know that P can only lie on a line as shown in Figure 4. If we also know the laser beam projection line relative to the local projector coordinate system and the 6 DoF (degrees of freedom) transformation between the projector and the camera, then we can figure out the laser beam projection line relative to the camera coordinate system. By intersecting the two lines it is possible to determine the coordinates of point P. In summary, the following parameters must be determined in a preliminary calibration step:

- intrinsic camera calibration matrix;
- 6 DOF transformation between the projector and the camera.
- Projection line relative to laser projector coordinate system. This is
- analogous to the intrinsic camera calibration matrix.

To get an accurate estimate of P, its pixel coordinates in the camera image must be known precisely. It can be shown that the wider the baseline the less effect errors in pixel coordinates have on the estimate of P. However, the baseline cannot be made very large because then the laser projector and the camera would have lesser overlapping field of view (FoV) and the laser spot may not be captured in the camera image at all.

The active triangulation method was invented to solve the notorious correspondence problem. The correspondence problem can be stated as follows: We are given two images of a scene captured from two different viewpoints. Let a 3D point P have pixel coordinates p in the first image. By only knowing the value of p - the coordinates of P, the intrinsic camera calibration matrix, and the 6 DoF transformation between the two camera viewpoints being unknown - find the image coordinates of P in the second image. Thus the correspondence problem consists of establishing which point in one image corresponds to which point in another, both being images of same point in space. Using laser light simplifies the problem drastically, since the P is "optically marked" with the colour of the laser light, so that it can be easily matched between the two images. An undesirable side effect is that the true colour at pixel p is lost, leaving a hole in the image, although holes the size of a pixel can be easily filled by interpolating the colour of surrounding pixels. Another issue is that if a red laser light is being used it will not reflect off, e.g., blue surfaces. Transparent objects such as glass also do not give off any reflection of laser light.

In order to determine depth of multiple points in the scene in a single step, a common solution is to use a scanning system, e.g. to sweep the laser beam by means of rotating mirrors while temporally synchronizing the camera and the laser projector with each other as the mirrors rotate. In its simplest form the active triangulation method gives one depth value per image i.e. 30 depth points per second if we use a camera with a 30 Hz frame rate. This is an extremely low rate, especially considering the amount of image data that is being captured, if we want to reconstruct a range map of the whole scene. There are various modifications of the technique such as projecting multiple points, lines, structured stripes and patterns, temporal coding, spatial coding, using IR projector and camera to get away with the problem of holes etc. giving rise to a plethora of new issues.

Conventional light sources can and are indeed used for optical triangulation as well, but laser sources have unique advantages for 3-D imaging. One of these is brightness, which cannot be obtained by an incoherent emitter. Another is the spatial coherence that enables the laser beam to "stay in focus" when projected on the scene. Nevertheless, this property is limited by the law of diffraction, which is written here as the propagation along the z axis of Gaussian laser beams:

$$\omega(z) = \omega_0 \left[1 + \left(\frac{\lambda z}{\pi \omega_0^2} \right)^2 \right]^{\frac{1}{2}}$$

Defining depth of field D_f by using the Rayleigh criterion gives:

$$D_f = 2 \pi \omega_0^2 / \lambda$$

which shows that the depth of field D_f is larger when the laser wavelength λ is small (toward the blue) and/or when the laser beam spot size ω_0 is large. This formula enables to estimate the image resolution of a diffraction-limited optical triangulation laser scanner and can be used as a guide when designing an optical system for imaging shapes, by showing the maximum number of pixels that can be acquired from a specific volume. For example, for a cube of 50 cm edge length, one has access to more than 2000 resolved spots in each direction axis X, Y, and Z using visible wavelength laser light, which correspond to a cubic element (voxel) of 200 μm on the surface of the illuminated object.

On scattering, the spatial coherence of the laser light is lost, which means that the depth of field used at the projection can be useful only if the lens aperture is closed down at the collection. Otherwise the focused laser spot is imaged as a blurred disk of light on the photodetector. A solution to this problem is to modify the conventional imaging geometry to conform to the Scheimpflug condition.

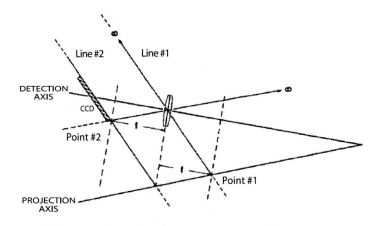

Figure 5. Geometric construction of the Scheimpflug condition.

Essentially this geometry enables the photodetector surface to "stay in focus" with the projected laser light. Its construction is very simple when the focal planes of the lens are used. Indeed, it is known that a point on the projection axis located at f (point 1) will be imaged at infinity (see Figure 5). Consequently the inclination angle of the photodetector is defined by drawing a line between that point and the principal point of the collecting lens (line 1). Similarly, on the other side of the lens, one knows that a point at infinity will be imaged at a distance f from the lens. The inclination and position of the photodector are then obtained by constructing a line (line 2) parallel to line 1 passing through the point 2.

Because of the coherent nature of lasers light, the spot imaged on the photodetector is corrupted with speckle noise [Baribeau 1991]. Figure 6 shows how speckle noise adds to the uncertainty of an imaged laser spot. The origin of the modulation observed in the profile is related to the scattering of a pure wavelength when diffusion occurs at the surface of the object.

Within the projected laser spot, each scatterer can be regarded as a coherent emitter, and because there are many of them within the area of illumination, the resulting image is the coherent sum of spatially incoherent sources. The result is a random modulation multiplying the expected smooth light profile. If σ^2 denotes the position variance which reflects the uncertainty in position of imaged laser spot then $\sigma^2 \sim \lambda$, with the proportionality factor depending on the geometry of the optical system. For a typical geometric set up, the centroid uncertainty is found to be of the order of a few micrometers, and more interesting, the physical dimensions of the photosensor elements have no effect on that limit. Consequently, the larger the photosensors, the better are the performances of the 3D digitizer in terms of sensitivity, speed, and depth of view.

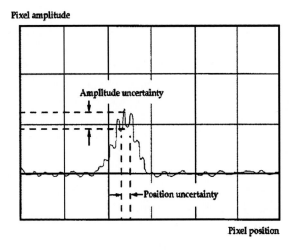

Figure 6. Speckle noise limits the position sensing resolution in triangulation LRF's.

AMPLITUDE MODULATION LASER RANGEFINDERS

The functioning principle of amplitude-modulated laser rangefinders is based on the indirect determination of the round trip time delay of the AM sounding beam through the measurement of the phase delay $\Delta\varphi$ of the signal photocurrent with respect to a reference signal. Distance is simply determined by the formula:

$$d = \frac{c}{n} \frac{\Delta\varphi}{4\pi v_m}$$

where v_m is the modulation frequency, c the velocity of light in vacuum and n the refraction index of the transmitting medium. For laser optical powers such that the signal shot-noise dominates over all other noise sources in the detection process, the accuracy of measurements can be showed to increase with the modulation frequency v_m, according to the formula [Nitzan 1977]:

$$\sigma_R \propto \frac{1}{m v_m (SNR)_i} \qquad (SNR)_i = \sqrt{\frac{P_S \eta \tau}{(h v)\Gamma}}$$

where σ_R is the "intrinsic" error (i.e. the minimum attainable error in optimal experimental conditions), m is the modulation depth and $(SNR)_i$ is the current signal-to-noise ratio - depending on the laser frequency v and collected power P_S, the integration time τ, the detector's quantum efficiency η and the overall optics merit factor Γ. All measured distances are relative to the sensor's position – specifically, to the center of the scanning mirror.

Because of phase 2π periodicity, AM range finding techniques are generally affected by the so-called "folding" ambiguity. In fact, using a single modulation frequency only enables to perform univocal distance measurements within a well-determined range window, equal to half the value of the corresponding modulation wavelength λ_m (e.g. ~15 m at 10 MHz). The position of points located in the scene out of this range is erroneously "folded", i.e. reported as if it was falling within the range. This is due to the periodicity of the modulation signal, which only permits the determination of distances up to multiples of $\lambda_m/2$. What is normally done in order to overcome this problem is modulating the laser light amplitude at two different frequencies, whose values are usually chosen far apart from each other. The low-frequency measurement range is chosen large enough to encompass the whole scene of interest. Low-frequency – unambiguous, yet less accurate – measurements are then used to remove the "folding" ambiguity that affects the corresponding high-frequency - more accurate – measurements. By using this double modulation technique, it is possible to measure distances at the level of accuracy permitted by the high-frequency mode but well beyond its intrinsic range. The method can profitably be used to digitize large scenes such as building façades or interiors, vaults etc [Ferri 2003] [Bartolini 2005].

The simpler single-modulation technique can also be successfully applied whenever the target is completely contained within the range corresponding to the chosen operating modulation frequency. The single-modulation technique is particularly well suited to the digitization of small-size objects, since a relatively high frequency can be chosen in this case without incurring in folding ambiguities.

A PROTOTYPE SYSTEM: THE IMAGING TOPOLOGICAL RADAR

The acronym ITR (Imaging Topological Radar) identifies a class of prototypal laser rangefinders designed and realised in the last years at the ENEA "Artificial Vision" laboratory in Frascati (Italy) [Bordone 2003] [Fantoni 2003][Ricci 2005]. ITR systems are based on amplitude-modulation range finding techniques, which enable to produce in a single scan an accurate range image and a shade-free, high resolution, photographic-like reflectivity image of the target under examination. These data can be used to produce faithful 3D digital models of real targets - either individual objects or complex scenes. Thanks to the pixel-by-pixel correspondence of the two images, reflectivity data are exploited in 3D rendering as grey-scale vertex colour information, resulting in highly realistic models. ITR systems provide submillimetric range measures (~100 μm) at distances that can vary from some tenths of centimetres to several tenths of meters. They are intrinsically non-invasive, due to the very low laser power (a few mW) required for their operation, and suited to be used in extreme or even hostile environments (high temperatures, radioactivity), because of the modular design that enables to keep the passive optical head separate from active module. The active module includes the laser source and the detector, while the passive module only contains the transmitting and receiving optics. The two subsystems are optically coupled trough optimized optical fibre connections. This modular setup enables to place the passive module in the position that is more convenient for the measure, without compromising the overall functioning of the system even in extreme or hostile conditions, such as at high or low temperature, and in presence of intense ionizing radiation background. A detail of the ITR optical head is shown in Figure 7.

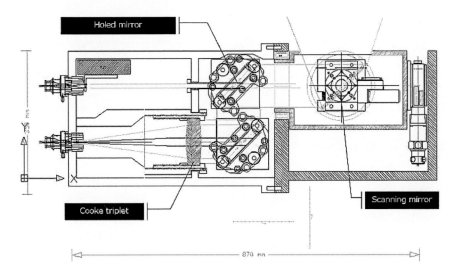

Figure 7. Scheme of ITR optical head.

3D Model Reconstruction with the ITR

Before actual 3D rendering, "raw" information contained in the pair of range and intensity 2D maps must be pre-processed in order to:

- convert range data values into Cartesian coordinates expressed in physical units and rearrange the geometrical information contained therein so as to get rid of the apparent distortions introduced by the viewing frustum of the scanning system (since the frustum is shaped like a spherical cap in the ITR case, we refer to the rearrangement procedure as "polar reformatting");
- remove extraneous elements such as background points, outliers, fake points that happen to be erroneously recorded by the sensor in correspondence with abrupt depth discontinuities.

In the rendering process intensity data are naturally used as vertex colour information, which enables to obtain an extremely realistic 3D model of the original object or scene. In cases when a series of linear scans are taken from different orientations, the corresponding partial range surfaces must be registered [Besl 1992] in a single frame. The last step of the rendering process consists in merging the registered range surfaces into a single polygonal model, possibly taking advantage of data redundancy in overlapping regions. Many algorithmic procedures have been devised at this purpose, such as surface zippering [Turk 1994] and volumetric methods [Curless 1996].

Although many software products exist on the market that supply general purpose tools for the rendering of most commercial 3D scanner data, they are generally not versatile enough to cope with all the requirements of a laboratory instrument like the ITR. For this reason we started implementing from the outset custom versions of the advanced algorithms needed for the 3D rendering of the intensity and range map pairs produced by the ITR. These tools are bundled in the "Isis" software package, a Windows application written in Microsoft Visual C++ with OpenGL support that comprises a control and acquisition tool and an interactive 3D reconstruction and visualization tool (see Figure 8).

The reconstructed models can be inspected at any angle by performing simple mouse-driven operations and exported in the most important 3D data formats. Current software development activities are aimed at implementing new advanced features, such as exploiting "colour" – i.e. intensity – information for enhancing the registration of multiple range surfaces. Colour-enhanced registration is expected to provide much better results in all situations where geometric features do not provide enough information to permit the convergence of the registration method (quasi-planar surfaces etc.). It is worth remarking that the simultaneous generation of distance and intensity information is a distinctive feature of AM range finders.

Very promising indications have been provided recently that the AM technique can be successfully applied also in underwater environments [De Dominicis 2005].

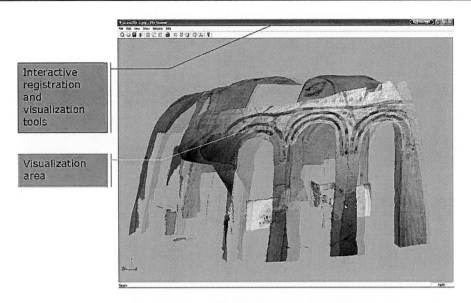

Figure 8. A screenshot of the Isis interactive 3D reconstruction and visualization tool.

APPLICATIONS TO THE RECONSTRUCTION OF UNDERGROUND CAVITIES

Among underground industrial applications, 3D laser scanning is widely adopted in the mining environment as a safer and more precise alternative to traditional surveys. Many specialised 3D laser scanning hardware, software and survey techniques have been developed to provide 3D survey services to the mining industry, with benefits that include:

- detailed quarry surveys, anytime of the year and in any conditions;
- full quarry modelling – including contouring creation and edge digitisation;
- remote and detailed modelling of underground stopes;
- pre and post blast profiles to determine blast efficiencies, i.e. the amount of displaced rock.

Common examples (see e.g. [Lester Franks]) of mining applications for 3D laser scanning include:

- accurate stockpile volumes and detailed profiles;
- periodic auditing of quarry volumes over the whole quarry site;
- profiling of pre and post blast faces;
- plan creation of bench or high-wall faces;
- digitisation of fault and interface lines;
- slope stability and deformation measurements;
- open stope measurements.

A typical commercial equipment used in this context is for example the Riegl LMS-Z210 terrestrial laser scanner, characterised by:

- data capture rate up to 6,000 points per second;
- 340° field of view (horizontal), ±40° field of view (vertical);
- 300m range
- Maximum point density of 1 point per 125mm at a range of 100m
- Single point precision of ±25mm

Another important field of application for 3D digitisation methods and technologies is given by cultural heritage archaeological excavations and 3D cataloguing and preservation of hypogeal tombs and churches.

The accurate recording of ancient crypt and grotto sites is a challenging task. These sites have either formed naturally or been carved from the surrounding rock and typically the walls, floors and ceilings have an irregular surface shape and paintings follow the contour of the rock surface over large areas. These features and particularly the shape of the rock surface are difficult to record, measure, compare and display by using conventional photographic and conservation recording techniques. LRF techniques are particularly well suited in this case, since they enable to produce accurate 3D models containing a wealth of information that can be examined and analysed for a number of conservation, research and conservation applications. For example, in the case of a site subject to limited access for conservation reasons, an immersive 3D virtual reality environment can be created to enable visitors to "virtually" visit the site, or experts to perform complex analyses without physically moving from their office. Researchers can for example magnify - zoom in - a 3D model up to the acquisition resolution limit to inspect and perform measurements on fine surface details in search for signs of deterioration or to examine brush stroke features. Computer-based visual enhancement and image processing techniques can be applied to accomplish "virtual restoration" tasks or to improve the intelligibility of faded images or inscriptions, as well as to remove graffiti. Last but not least, 3D models recorded before conservation treatments can be easily archived and used for cataloguing, monitoring and maintenance purposes.

In the specific context just outlined, many examples of successful applications can be reported. We limit ourselves to cite [Fiorani 2000], which describes the application of GEOLIDAR, a miniaturized LRF prototype, to the volumetric characterizations of underground cavities, such as buried tanks, temples, and tombs. Specifically, the authors used a coring machine to drill a small-diameter hole up to the cavity where GEOLIDAR is let down and execute a motor-driven 3D scan.

Because of its characteristics of robustness and adaptability to extreme or even hostile conditions, the ITR, described in detail in the previous section, is also very well suited to 3D digitisation applications in underground environments, especially in the context of archaeological underground investigations and diagnosis of hypogeal structures such as tombs and churches.

We report in the following a recent example of 3D reconstructions carried out by using the ITR for data acquisition and accompanying software for pre-processing and 3D rendering.

"Grotta dei Cervi", Porto Badisco (Lecce, Italy)

The Grotta dei Cervi (lit. grotto of deer) close to the village of Porto Badisco, Lecce, southeastern Italy, was discovered in 1970 by a group of speleologists, who called it "Grotta di Enea" on the basis of a local tradition. Later on it was named "Grotta dei Cervi" because of the presence of paintings showing deers. The Grotta dei Cervi is of karstic origin and contains the largest and most important set of paintings from the European Neolithic, thanks to the hundreds of pictures painted on the walls of its galleries and in the many chambers it is composed of. The archaeological findings show human dwellers started inhabiting the cave between the mid-Neolithic and the early Chalcolithic, when some corridors became obstructed. The paintings were created during this time span (about 4,000-3,000 b.c.), although it is difficult to determine their precise chronological sequence. They include many graphic elements (cross-shaped, comb-shaped, spiral-shaped, etc.) as well as figurative elements in the shape of human beings, dogs and deer, often giving rise to hunting scenes (see Figure 9).

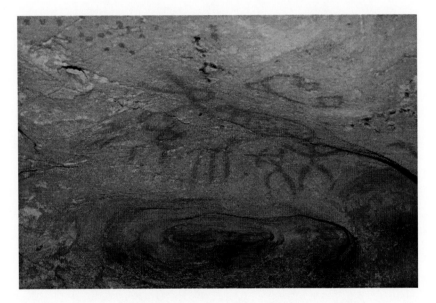

Figure 9. Photographic picture representing some prehistoric drawings in the Grotta dei Cervi.

The paintings are especially frequent in the area that is closest to the entrance of the cave, whilst they become less frequent, and ultimately dwindle to nothing, as one moves towards the remotest parts of the secondary galleries. Here hand-shaped marks and abstract motifs predominate that are more difficult to interpret. The groups of paintings - mostly in brown from the guano they are made of and only exceptionally in red - are located along three corridors for a total length of about 400m. Their distribution shows that the different sections of the cave were intended for different purposes and testifies the use of the cave as a place of worship – the grotto is also known as "the shrine of Prehistory", according to an archaeologist's evocative definition.

Motivated by the interest of archaeologists in determining the exact vertical positioning of the paintings, which could be the key to revealing a "hidden magical meaning" and shed light on Neolithic religious rituals, a project for the 3D digitization of the grotto was launched in 2004 involving, among others, the ENEA "Artificial Vision" Laboratory and the Soprintendenza Archeologica di Lecce – the "Soprintendenze Archeologiche" are decentralised organs of the Italian "Ministero per i Beni e le Attività Culturali" with responsibilities for the safeguarding and enhancement of the cultural heritage.

The 3D digitization of the wall paintings and hypogeal environments of the Grotta dei Cervi, hardly accessible and characterized by high humidity, was a complex task, requiring high-tech instruments suited to use in extreme conditions such as the ITR. The acquisition campaign lasted several days and was carried out by introducing only the ITR passive module in the cave main gallery, while keeping active instrumentation and the control module outside, at several tenths of meters, to preserve it from moisture (see Figure 10).

Figure 10. Details of the acquisition campaign in the Grotta dei Cervi: the ITR passive module (on the left) is transported in the cave along narrow rock tunnels (on the right).

After registering and fusing range images acquired from different viewpoints and at various resolutions (see Figure 11), a single 3D model of the grotto was finally realized (see Figure 12).

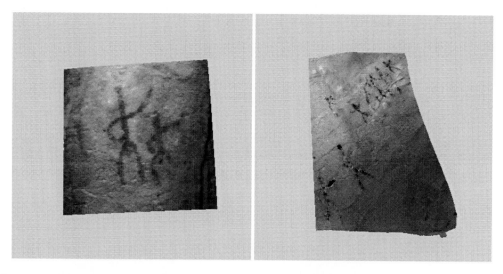

Figure 11. High resolution details of the 3D model of the "Grotta dei Cervi."

Figure 12. Full 3D reconstruction of the "Grotta dei Cervi" main gallery (just to give an idea of the linear size, the distance from the floor to the ceiling top is ~15m).

The final 3D model, consisting of several million triangles, was subsequently severely decimated by adopting an adaptive approach, i.e. by preserving the highest accuracy in those parts where paintings were present while reducing the overall model complexity.

CONCLUSION

We presented a broad survey of laser rangefinding methods and techniques. Applications of LRFs were briefly discussed and the most important methods of laser ranging were described, including interferometric, active triangulation and time of flight methods. Salient features of these methods were noted and limitations were mentioned. One important technique, namely amplitude-modulation (AM) rangefinding, was covered in greater detail because of its relevant applications in the specific context of 3D digitization of underground environments, such as caves and archaeological excavations. Recent progress in the development of AM LRFs suited to be operated underwater in real conditions [De Dominicis 2005] makes this technique very promising also for undersea archaeology.

REFERENCES

Amann, M.C., Bosch, T., Lescure, M., et al. (2001) "Laser ranging: a critical review of usual techniques for distance measurement", *Optical Engineering,* 40(1):10–19.

Baribeau, R. & Rioux, M. (1991) "Influence of speckle on laser range finders", *Applied Optics*, 30(20):2873–2878, 1991.

Bender, P.L. (1967) "Laser Measurements of Long Distances", *Proceed. of the IEEE*, Vol. 55(6), pp. 1039-1045.

Besl, P. & McKay, N. (1992) "A Method for Registration of 3-D Shapes", Trans. PAMI, Vol. 14, No. 2.

Bien, F., Camac, M., Caulfield, H.J., et al. (1981) "Absolute distance measurements by variable wavelength interferometry," *Applied Optics*, Vol. 20(3), pp.400-403.

Bartolini, L., Ferri De Collibus, M., Fantoni, R., et al. (2005). "Amplitude-modulated laser range-finder for 3D imaging with multi-sensor data integration capabilities", *Proceed. of SPIE*, Vol. 5850 *Advanced Laser Technologies,* 2004, 152-159.

Bordone, A., Ferri De Collibus, M., Fantoni, R., et al. (2003). "Development of a high-resolution laser radar for 3D imaging in artwork cataloguing," *Proceed. of SPIE*, Volume 5131, 244-248.

Bufton, J.L. (1989) "Laser Altimetry Measurements from Aircraft and Spacecraft", *Proceed. of the IEEE*, Vo. 77, No. 3, March 1989.

Curless, B. & Levoy, M. (1996) "A volumetric method for building complex models from range images", *Proceed. of SIGGRAPH*, ACM, pp. 303-312.

De Dominicis, L., Bartolini, L., Ferri De Collibus, M., et al. (2005). "Underwater three-dimensional imaging with an amplitude-modulated laser radar at a 405 nm wavelength", *Applied Optics;* Vol. 44 No. 33, 1-6.

Fantoni, R., Bordone, A., Ferri De Collibus, M., et al. (2003). "High resolution laser radar: a powerful tool for 3D imaging with potential applications in artwork restoration and medical prosthesis", *Proceed. of SPIE,* Vol. 5147 ALT'02, 116-127.

Ferri De Collibus, M., Fantoni, R., Fornetti, G., et al. (2003). *High-resolution laser radar for 3D imaging in artwork cataloguing, reproduction and restoration, Proceed. of SPIE,* Vol. 5146, 62-73.

Fiorani, L., Bortone, M., Mattei, S., et al. (2000) "Miniaturized laser range finder for the volumetric characterization of underground cavities", *Proceed. of SPIE*, Volume 4129, Subsurface Sensing Technologies and Applications II, Cam Nguyen, Editor, July 2000, pp. 457-463.

Goldstein, B.S. & Dalrymple, G.F. (1967) "Gallium Arsenide Injection Laser Radar" *Proceed. of the IEEE*, Vol. 55 (2), pp. 181-188.

Hovanessian S.A. (1988) "Introduction to Sensor Systems", *Artech House,* p.2.

Lester Franks Ltd, (http://www.lesterfranks.com.au).

Lyöri, V., Määttä, K., Nissilä, S. et al. (1997) "A high precision Fresnel-OTDR for distributed fibre-optic sensor network applications", 12th International Conference on Optical Fibre Sensors, Oct 28-31, 1997, USA, pp. 520-523.

Ma, Y., Soatto, S., Kosecka, J., et al. (2003) "An Invitation to 3-D Vision: From Images to Geometric Models", chapter 3, Springer-Verlag, 2003.

Nayar, S., Watanabe, M. & Noguchi, M. (1995) "Real-time focus range sensor, *Proceed. of IEEE International Conference on Computer Vision*, pp. 995–1001.

Nitzan, D., Brain, A.E. & Duda, R.O. (1977) "The Measurement and Use of Registered Reflectance and Range Data in Scene Analysis", Proc. IEEE 65, 206.

Ricci, R., Ferri De Collibus, M., Fantoni, R., et al. (2005). "ITR: an AM laser range finding system for 3D imaging and multi-sensor data integration", *Proceed. of ICST 2005* (International Conference on Sensing Technology, Palmerston North, New Zealand, 21-23 November 2005), 641-646.

Riegl USA Inc. (2001) Application sheets (http://www.rieglusa.com/).

Turk, G. & Levoy, M. (1994) "Zippered polygon meshes from range images", *Proceed. of SIGGRAPH*, ACM, pp. 311-318.

Vali, V, Krogstadt, R.S., Moss, R.W., et al. (1966) "Measurement of strain rate as the Kern River using a laser interferometer", Trans. Am. Geophys. Union, vol. 47, June 1966, p. 424.

INDEX

2

2D, 151, 228, 234, 236, 242

3

3D-model, viii

A

Aβ, 133, 172
absorption coefficient, 52, 54, 132, 173, 216
absorption lidar, viii, 4, 71, 73, 74, 77, 78, 79, 81, 85, 88, 89, 91
absorption spectra, 30, 55, 56
access, 123, 160, 238, 244
accounting, 101
accuracy, 80, 91, 104, 105, 106, 107, 109, 111, 112, 113, 124, 125, 126, 127, 128, 157, 169, 174, 175, 182, 187, 188, 192, 195, 197, 227, 230, 231, 233, 235, 240, 247
ACE, 78, 84
acid, 15, 17, 73, 212
ACM, 248, 249
acoustic waves, 101
activation, 195
adaptability, 244
ADC, 137
administrators, 114
AE, 129, 151
aerosols, vii, 17, 34, 60, 67, 74, 75, 77, 78, 79, 81, 83, 87, 88, 90, 138, 160, 183, 187, 221
Ag, 175, 177, 178, 179, 180
age, 68, 147, 148
agent, 17
aggregates, 138
aggregation, 36

agriculture, 16, 195
aid, 111
aircraft, 3, 59, 60, 61, 62, 63, 64, 78, 83, 95, 96, 98, 101, 102, 103, 104, 105, 106, 124, 127
Alabama, 116, 118
algae, 210
algal, 138, 139, 140
algorithm, 34, 35, 36, 61, 67, 123, 124, 145, 152, 197, 208
alloys, 178, 179, 192, 221
alpha, 224
Alps, 84
ALT, 248
alternative, 47, 54, 56, 243
aluminum, 148, 192, 202, 208, 210, 211
ambiguity, 240
amino acid(s), 138
amplitude, viii, 26, 100, 102, 136, 151, 227, 228, 230, 232, 234, 239, 240, 241, 247, 248
Amsterdam, 221
analytical techniques, 195, 197
angular momentum, 26
anisotropic, 151
anisotropy, 19, 26
Antarctic, viii, 63, 84, 88, 144, 147, 148, 149, 152, 198, 202, 203, 208, 211, 222
antenna, 4
anthropogenic, vii, 16, 17, 46, 107, 143, 146
aquatic, 139, 143, 195
aquatic systems, 143
Arctic, 17, 61, 74, 79, 82, 84, 85, 90
argon, 202, 225
Army Corps of Engineers, 98, 114, 116, 127
aromatic, 146
aromatic compounds, 146
artistic, 68, 166
ash, 212, 216
ASI, 153, 220
Asian, 84
aspect ratio, 185

B

C

D

E

J

K

L

Q

R